Agricultural Extension System
The Way Forward

W0234614

Agricultural Extension System
The Way Forward

Sarthak Chowdhury MSc (Ag), PhD
Professor and *Former* Principal
Department of Agricultural Extension
Palli Siksha Bhavana (Institute of Agriculture)
Visva-Bharati, Sriniketan, West Bengal

Debabrata Mondal MSc (Ag), PhD
Assistant Teacher, Mahesh High School
Mahesh, Hooghly, West Bengal

OXFORD
& IBH

Oxfort & IBH Publishing Co. Pvt Ltd
New Delhi (A Unit of CBS Publishers & Ditributors Pvt Ltd)

CBS

CBS Publishers & Distributors Pvt Ltd
New Delhi • Bengaluru • Chennai • Kochi • Kolkata • Mumbai
Hyderabad • Jharkhand • Nagpur • Patna • Pune • Uttarakhand

Agricultural Extension System
The Way Forward

ISBN: 978-93-90709-91-5

Copyright © Authors and Publisher

First Edition: 2021

OXFORD & IBH
New Delhi
(*A Unit of* CBS Publishers & Distributors Pvt Ltd)

Published by **Satish Kumar Jain** and produced by **Varun Jain** for

CBS Publishers & Distributors Pvt Ltd
4819/XI Prahlad Street, 24 Ansari Road, Daryaganj, New Delhi 110 002, India.
Ph: 011-23289259, 23266861, 23266867   Website: www.cbspd.com
Fax: 011-23243014   e-mail: delhi@cbspd.com; cbspubs@airtelmail.in.

Corporate Office: 204 FIE, Industrial Area, Patparganj, Delhi 110 092
Ph: 011-4934 4934   Fax: 011-4934 4935   e-mail: publishing@cbspd.com; publicity@cbspd.com

Branches

- **Bengaluru:** Seema House 2975, 17th Cross, K.R. Road, Banasankari 2nd Stage, Bengaluru 560 070, Karnataka
  Ph: +91-80-26771678/79   Fax: +91-80-26771680   e-mail: bangalore@cbspd.com
- **Chennai:** 7, Subbaraya Street, Shenoy Nagar, Chennai 600 030, Tamil Nadu
  Ph: +91-44-26680620, 26681266   Fax: +91-44-42032115   e-mail: chennai@cbspd.com
- **Kochi:** 42/1325, 1326, Power House Road, Opp KSEB, Ernakulum, Kochi 682 018, Kerala, India
  Ph: +91-484-4059061-65,67   Fax: +91-484-4059065   e-mail: kochi@cbspd.com
- **Kolkata:** 6/B, Ground Floor, Rameswar Shaw Road, Kolkata-700014 (West Bengal), India
  Ph: +91-33-2289-1126, 2289-1127, 2289-1128   e-mail: kolkata@cbspd.com
- **Mumbai:** PWD Shed, Gala no 25/26, Ramchandra Bhatt Marg, Next to JJ Hospital Gate no. 2, Opp. Union Bank of India, Noorbaug Mumbai-400009, Maharashtra
  Ph: +91-22-66661880/89   e-mail: mumbai@cbspd.com

Representatives

• Hyderabad	0-9885175004	• Jharkhand	0-9811541605	• Nagpur	0-9421945513
• Patna	0-9334159340	• Pune	0-9623451994	• Uttarakhand	0-9716462459

Printed at Rashtriya Printers, Dilshad Garden, Delhi, India

Preface

It goes without saying that extension as an institution is the only one component in agriculture and rural development process and that is only one vehicle for fostering change in agriculture and rural development. In the past two decades agricultural extension services in developing countries have been under increasing pressure from globalization, liberalization of agricultural markets, environmental changes and food insecurity.

Agricultural extension in India has undergone several changes since independence. Still, a large number of resource poor farmers and other vulnerable groups remain unreached by the public extension system. Present day agricultural extension faces the challenges in the area of relevance, accountability and sustainability. Several efforts have been made in the public sector over the past one decade to initiate various reform measures and operational models to improve the organizational performance of the system. Yet, the challenge of enhancing relevance, efficiency and effectiveness of the public sector agricultural extension system in meeting its organizational goals and objectives remain unresolved.

Keeping this in mind this book analyses Indian major reform initiative implemented to create a demand-driven, broad-based and holistic agricultural extension system. In addition, the book evaluates the effects of the agricultural sector reform initiatives on access to and the quality of agricultural extension service. The ultimate objective is to gain view on what works, where and why in improving the effectiveness of India's agricultural extension system, to identify measures that strengthen and improve agricultural extension service provision, and to reveal existing knowledge gap. The ethics of extension education, criticism over present forms of extension have been synthesized to provide guidelines for students, researchers and practitioners.

The students, academicians, research scholars and scientists will find this book very useful. In addition, Agricultural Extension Officers, Agricultural Development Officers of the line departments of the states as well as the Subject Matter Specialists (SMSs) of Krishi Vigyan Kendra (KVK) and practitioners would find this book very useful.

Sarthak Chowdhury
Debabrata Mondal

Contents

List of Acronyms

AC & ABC	Agri-Clinics and Agri-Business Centres
AEO	Agricultural Extension Officer
AES	Agricultural Extension System
AICRP	All India Coordinated Research Project
AIDS	Acquired Immunodeficiency Syndrome
AIS	Agricultural Innovation Systems
AKIS	Agriculture Knowledge and Information System
AMC	ATMA Management Committee
AT	Appropriate Technology
ATIC	Agricultural Technology Information Centre
ATMA	Agricultural Technology Management Agency
BAP	Block Action Plans
BARC	Bhabha Atomic Research Centre
BTT	Block Technology Team
CAFE	Capacity Building of Farmers in Agriculture for Farmer led Extension
CAU	Central Agricultural University
CSIR	Council of Scientific and Industrial Research
DAC	Department of Agriculture and Cooperation
DACFW	Department of Agriculture and Farmers' Welfare
DAESI	Diploma in Agricultural Extension Services for Input Dealers
DARE	Department of Agricultural Research and Education
DFID	Department for International Development
DoE	Directorate of Extension
DRWA	Directorate of Research on Women in Agriculture
EEI	Extension Education Institute
FAO	Food and Agriculture Organization
FFS	Farmer Field School
FIAC	Farm Information and Advisory Centre
FIG	Farmers Interest Group
FLD	Front Line Demonstration
FO	Farmers Organizations
GB	Governing Board
GDP	Gross Domestic Product
HIV	Human Immunodeficiency Virus

HRD	Human Resource Development
IAAP	Intensive Agricultural Areas Programme
IADP	Intensive Agricultural District Programme
ICAR	Indian Council of Agricultural Research
ICT	Information and communications technology
IFFCO	Indian Farmers Fertilizer Cooperative Limited
INM	Integrated Nutrient Management
IPM	Integrated Pest Management
IQ	Intelligence Quotient
ITD	Innovations in Technology Dissemination
IVLP	Institution Village Linkage Programme
KCC	Kisan Call Centre
KGK	Krishi Gyan Kendras
KPS	Krishi Prajukti Sahayak
KVK	Krishi Vigyan Kendra
LCD	Liquid Crystal Display
MANAGE	National Institute of Agricultural Extension Management
MC	Management Committee
MM & ICT	Mass Media and Information & Communication Technology
MoA	Ministry of Agriculture
NAARM	National Academy of Agricultural Research Management
NABARD	National Bank for Agriculture and Rural Development
NAEP	National Agricultural Extension Project
NAFED	National Agricultural Cooperative Marketing Federation of India
NARP	National Agricultural Research Project
NARS	National Agricultural Research System
NATP	National Agricultural Technology Project
NDDB	National Dairy Development Board
NGO	Non-Government Organisations
NGRCA	National Gender Resource Center in Agriculture
NMAET	National Mission on Agriculture Extension and Technology
NRC	National Research Centres
NRM	Natural Resource Management
NRSA	National Remote Sensing Agency
OFAI	Organic Farming Association of India
OFT	On-Farm Testing
OHP	Over Head Projector
PFAE	Policy Framework for Agricultural Extension
PPP	Public-Private Partnership

PRA	Participatory Rural Appraisal
PRI	Panchayati Raj Institutions
PSU	Public Sector Undertaking
R & D	Research and Development
R-E	Research and Extension
RRA	Rapid Rural Appraisal
SAMETI	State Agricultural Management and Extension Training Institute
SAU	State Agricultural University
SHG	Self Help Group
SMS	Subject Matter Specialist
SREP	Strategic Research and Extension Plan
T & V	Training and Visit
TAR	Technology Assessment and Refinement
TDMC	Technology Dissemination Management Committee
TOC	Theory of Constraint
TOT	Transfer of Technology
TTC	Trainers' Training Centres
TV	Television
UGC	University Grants Commission
UNEP	United Nations Environment Programme
USA	United States of America
VEW	Village Extension Worker
WB	The World Bank
WGAE	Working Group on Agricultural Extension
ZPD	Zonal Project Directorate
ZRS	Zonal Research Station

Changing Scenario in Agricultural Extension

Poverty, hunger, economic growth, food production, and natural resource degradation are the major challenges in today's world. Ensuring a thriving agricultural economy is critical for reducing poverty, enabling food security and managing natural resources in a sustainable fashion. Agriculture has already reached the limits of land and water, thus future increases in food production must exploit biological yields on existing land. As the world grapple with these issues, agricultural extension faces the challenges in the area of relevance, accountability and sustainability.

Inspite the impressive progress that India has made since independence in the field of science and technology, the rural sector and people remain grossly underdeveloped. The sector is characterized, inter alia, by the preponderance of small and scattered rural enterprises, lack of basic infrastructure, low productivity, excessive dependence on weather and climatic factors and consequently high degree of risk and uncertainty (Singh, 2009).

Geographical location, food insecurity, and being poorly served by development departments, education, heath, transport, communication and other services characterize the rural environment in most developing countries. Agricultural productivity and its associated agricultural extension services are important to the livelihood activities of rural communities. For rural communities to fullfil their respective needs require access to productive services and information on input supply, new technologies; credit, and market. Agricultural extension service has been identified as an important part of the intended transformation of the agricultural sector.

The shifting emphasis of Indian agriculture towards diversification, commercialization, sustainability and efficiency has made it necessary for the state extension organizations, to critically examine their extension approaches. Department of Agriculture in several states made changes in some of their approaches towards the late 1980's. But the basic issues regarding the type of support required by farmers and the changes in extension organization needed to provide these were not addressed (Van den Ban and Sulaiman, 2000).

It goes without saying that extension as an institution is the only one component in agriculture and rural development process and that is the only one vehicle for fostering change in agriculture and rural development. Agricultural extension over the years has been used as tool for facilitating agricultural and rural development (Chambers,

1997; Alex and Byerlee, 2002). Extension organizations therefore, play an important role in rural development in developing countries (Shackleton et al., 2000; Mwabu and Thorbecke, 2001).

The world has entered a new economic system that has evolved from structural adjustment and trade liberalization; technological progress; advances in telecommunication and greater interdependence of world labour, product and financial market. In the past two decades agricultural extension services in developing countries have been under increasing pressure from globalization, liberalization of agricultural markets, environmental changes, HIV/AIDS and food insecurity, to reform and respond to the need of their client (World Bank, 2000a; Richardson, 2003). Agricultural Extension in Asia, particularly in low income countries, is struggling to reinvent itself. For decades the policy emphasis has been on public sector provision of services to extend new technologies to farmers. Public extension has and will continue to play an important role in most Asian countries. Without public fund for extension, sustainable public interest are compromised especially those concerned ecological sustainability and poverty reduction (Katz, 2002).

In response to these changes, extension organizations are shifting their principal focus from agricultural productivity alone towards sustainable development, where participatory process, action learning, i.e. the human dimension of agricultural and rural resource management are given importance (Scoones and Thompson 1994; Van Crowde 1996b). Agriculture extension operates within a broader knowledge system that includes research and agricultural education. FAO and the World Bank referred to this larger system as Agricultural Knowledge and Information System (AKIS).

The growing consensus to create a demand driven technology system with direct involvement of farmers in identifying problems, establishing priorities, and carrying out on farm research and extension activities, need a farmer friendly institution or organization to strike a balance between institutional supply system and farmers initiated demand driven extension system.

The above observations provide the very logic for institutional reforms for a newer version for the agriculture and rural development realized much long back. Food and Agriculture Organization (FAO, 2001) of United Nations emphasized the need for institutional reform in agriculture and rural extension in developing countries keeping world in view. In due course, in an effort to respond to the new paradigm, countries worldwide had adopted a variety of institutional reforms. These reforms are either market oriented or non-market oriented (Smith, 1997).

The importance of agricultural extension in transferring relevant knowledge and information to farmers as well as in translating policy directions into action is well known. India has a long tradition of agricultural extension. Agricultural extension in the post-independence era was largely the function of State Departments of Agriculture. Some voluntary organisations were also involved in agricultural development activities in different parts of the country, but with limited outreach. The Indian Council of Agricultural Research (ICAR) began its participation in agricultural extension through National Demonstrations in 1964.

A major change in public sector extension came with the implementation of the World Bank sponsored Training and Visit System (T&V) in 1974. Most States adopted the T & V system during the 1980s, and this improved the financial and human resource capacity of the extension system. The 1970s also witnessed the launch of Krishi Vigyan Kendras (KVKs) or Farm Science Centres, Lab-to-Land programmes, and Operational Research Programmes by the ICAR. Krishi Vigyan Kendras (KVKs) were begun by ICAR to provide need-based and skill-oriented vocational training to farmers, field-level extension workers and other self-employed persons. KVKs were meant to bridge the gap between technology developed at research institutions and its adoption at the field level. Their role was to feed proven technologies to the main extension system. The KVK programme began in 1974. There are now a total of 642 KVKs in the country– 429 under State Agricultural Universities (SAUs) and Central Agricultural Universities (CAU), 56 under ICAR institutes, 100 under Non-Government Organisations (NGOs), 35 under State Governments, three under various Public Sector Undertakings (PSUs), and the remaining 18 under other educational institutions. KVKs work under the administrative control of Zonal Project Directorates (ZPDs). There are 8 ZPDs in the country. In 1992, National Demonstrations, Operational Research Projects, and the Lab-to-Land Programme were merged with KVKs, and front-line demonstrations and on-farm testing were added to the responsibilities of KVKs. From 2009 onwards, KVKs have also assumed the role of knowledge and resource centres in the concerned districts. Each KVK has scientific manpower of six to seven subject-matter specialists.

Low manpower resources restrict the reach of KVKs to a limited number of farmers. Many KVKs are constrained by financial, infrastructural, and human resource limitations and unable to reach the farming community of a district.

Agricultural extension witnessed a qualitative change in the 1990s, with a new focus on privatization and the withdrawal of support to the state-led extension system. Reduced spending by government weakened the public sector extension system. Other non-governmental agencies stepped into fill the vacuum.

Facing criticisms on the failure of extension, the government introduced the Agricultural Technology Management Agency (ATMA). The ATMA model was pilot-tested from 1998 to 2005 in 28 districts, and later extended to all 548 rural districts in the country. The ATMA model was meant to make the extension system a demand-driven, market-oriented, and farmer-accountable system. At the district level, ATMA was to function as a registered society of all major stakeholders in agriculture and allied activities, with the objective of becoming a platform for the convergence of the various agencies involved in extension in a district. ATMA was to be the nodal point at the district level for technology dissemination, integrating research and extension activities, and decentralizing day-to-day management of the public agricultural extension system. Field-level activities are coordinated through Farm Information and Advisory Centres (FIAC) at the block level. Another feature of ATMA is that it deals with groups such as farmer groups or self-help groups rather than with individuals for the delivery of extension services. It also has provisions for public-private partnership in the district. In 2000, ICAR introduced Agricultural Technology Information Centres (ATIC) in selected ICAR institutes and State Agricultural Universities to function as a single window to disseminate technologies developed in the Universities and Institutes.

Many new service providers and institutional arrangements in agricultural extension have emerged over the last two decades. These include private extension agencies, input agencies, agri-business firms, farmers' organisations, producer cooperatives, financial agencies involved in rural credit delivery, and consultancy services (Sulaiman, 2012). The establishment of Agri-Clinics and Agri-Business Centres (AC & ABC) Scheme was an explicit move by government to support private sector initiatives in extension. Under the AC and ABC scheme, unemployed farm graduates were provided training for two months each and given access to credit to start their own ventures. Close to 45904 farm graduates were trained between 2002 and 2016 and more than 19402 ventures begun (AC and ABC 2016). The impact of this initiative is yet to be evaluated.

While the Indian extension system is now guided by a variety of models, schemes, and institutions, public sector extension continues to dominate. Though ICAR's extension initiatives have been important to transformations in Indian agriculture, their capacity and reach has always been limited compared to those of first-line extension systems run by State-level departments of agriculture. Further, since agriculture is a State subject, the mode of organisation and operation of public extension systems vary widely across States.

Indian agriculture is confronting serious issues such as a huge yield gap, a multitude of smallholders, imbalances with respect to input use and declining natural-resource productivity. Extension systems in India, which have an important role to play in addressing these concerns, are constrained by financial, infrastructural, and human resource limitations. An analysis of extension expenditure showed a serious setback in the 1990s. There is an immediate need to increase investment in extension.

The inclusiveness of extension services remains a major concern. Considering the prevalence of smallholders in Indian agriculture and the complexity of the problems confronting them, suitable extension strategies need to be formulated. The growth of smallholder agriculture will be determined by the extent to which institutions of research and extension are attuned to their priorities.

The focus of agricultural extension has been on increasing yield with much less attention paid to ecosystem health and natural resource conservation. Given the public-good nature of many of the benefits of natural-resource management activities, the role of government is critical.

Lastly, while there are a variety of institutions in the field of extension, the ability of private extension to reach disadvantaged and marginalized areas, enterprises and sections of society is not yet established. While private and non-governmental institutions should be encouraged, public extension has to be strengthened to cater to the scale and diversity of agriculture in India.

The challenge for smallholder farmers in India is typical (Birner and Anderson, 2007; Chandrasekhar Rao et al., 2011; Reardon et al., 2011). These farmers tend to have minimum access to information. Reaching farmers who search for information the least, would, therefore, require different content, approach and delivery mechanisms, as they have different information needs and rely mostly on interpersonal sources.

It is noted that in addition to technology transfer, agriculture and rural extension is a unique service that provides access to small farmers and rural poor leaving away from urban environment. It can provide this population with service to increase their productivity. Food security will depend on institutional development and income generation together with increased food crop output. According to Sasakawa Global-2000 programme, the long term strategy to overcome world hunger lies in helping the poor to produce more and better quality staple food more efficiently in order to take first step out of poverty. All the view together implies the need to raise farm productivity per unit of input and improve the competitiveness of food marketing system. FAO (2001) emphasized that reducing poverty and food insecurity is not simply a question of enhancing agriculture productivity and production for generating more income. Institutions are the structuring features that command access of people to assets, to voice and to power over the lives and that regulates their competing claims to limited resources. The fundamental need is to address institutional governance and politico-economic factors that tend to exclude individual and population groups from progress.

There is an increasing recognition all over the world that institutions are fundamental to the economic change. Agricultural development depends on an efficient flow of information among all the actors in the system, and agricultural extension has been traditionally performing this role with varying level of success. Its important contribution in promoting agricultural development and increasing food production have resulted in increased interest in extension during the last few years (Van den Ban and Hawkins, 1998).

It may be mentioned that the types of extension reform being initiated are not necessarily new. The theme of reform being initiated during the last two decades revolved around decentralization, participation and linkage which have been adopted in the countries like Australia, Brazil, Canada, German, India and United States. What is new, however, is the extent of globalization and economic restructuring in both developed and developing countries. The extension reforms have been categorized under market reform and non-market reforms. The market reform encompasses four major reform strategies like (i) revision of public sector extension; (ii) pluralism; (iii) cost recovery and (iv) total privatization. The non-market reform comprises two major strategies like decentralization (transfer of authority to lower tier) and subsidiary delegating responsibility to the lower level of hierarchy.

At present three decentralization directions currently dominate development of agriculture and rural extension. One is to decentralize the burden of extension cost by redesigning the fiscal system and other is to decentralize the central government responsibility for extension through structural reform. Another type of decentralization has come to exist which intends programme management through farmer participatory involvement in programme planning and decision making and ultimately the farmer to take responsibility of the extension programme. The advantage and limitation in each case is location specific and policy specific.

However, decentralization is most often thought of as a shifting of authority for extension to the lower tier of government. In general it involves transfer of funding and management authority to lower level of hierarchy. The success of extension system reforms crucially depends on how the research system responds to meet the needs of extension reforms. The most important reform measure from ICAR that relates to the implementation of the extension reform was the issue of a set of directives jointly with the Department of Agriculture and Cooperation (DAC-DARE, 2011; ICAR, 2011) directly indicating efforts to strengthen research-extension linkage.

The FAO consultation on agriculture extension 1990 recommended that all national governments should develop and periodically review their agriculture extension policy.

The policy should include goals of extension, agencies and their personnel, clients to serve and the programmes to address the problems. The other emphasis was supported by Swanson (1990) who stated for a comprehensive policy on extension system to contribute to agricultural productivity, farm income and improved quality of live.

Qamar (2002) stated that in central Asia the countries which exercised socialistic policies for many years shifted their attention to market oriented economy. To exploit full potential in agriculture appropriate national agriculture system was established through institutional reform and backed by national policies.

India is no exception to the global trends of reform. It has implemented a number of reforms in public sector extension system (Planning Commission, 2002 and 2007). In the last decade public sector extension in India has gained significant focus in policy cycle because it is seen as the weakest link in the research-extension-farmer-market chain to increase the agriculture growth to a target of 4% per year (Parsai, 2010). In order to meet these challenges, India's extension system has experienced major changes since the late 1990s in governance structure, capacity, organization and management, and advisory methods. The change involve the decentralization of extension service provision to the local level, the adoption of pluralistic modes of extension service provision and financing, the use of participatory extension approaches, capacity training to farmers to express their demands, and capacity training of service providers to respond to he demands of farmers, among others (Rivera, Qamar and Van Crowder, 2001; Birner et al. 2006; Birner and Anderson 2007; Anderson 2007). Moreover according to Byrnes (2001), extension can most effectively carry out its mandate, not by working directly with individual farmers' groups or organizations. The very concept emphasized the need for structural reforms in farmers groups and organizations.

Public sector extension is a state responsibility in India that has undergone several transformations since independence. Initially, the focus of extension was on human and community development. But for food security there has been a gradual and focused shift in interest towards technology transfer within the policy framework. The most significant development during seventies was introduction of T & V extension management system; a system well suited to the rapid dissemination of broad based crop management practices for high yielding wheat and rice varieties released during mid-sixties. To support green revolution technology for major cereal crops, the extension activities has been given over emphasis and implemented through department of agriculture. The other departments like horticulture, animal husbandry and fisheries did not receive adequate attention but continued with provision of subsidized inputs and services to farmers. By the early nineties with the completion of third National Agricultural Extension Projects (NAEP), the important contribution that T and V extension system had made to agricultural development were duly recognized. Subsequently, realization was made about the challenge in meeting the need of the farmers in the 21st century.

The system approach was conceived to overcome the earlier constraints associated with different approaches. The location specific approach came to prominence as the need of the farmers in rain-fed and irrigated areas were quite diversified in respect of livestock, horticulture and high value commodities that are capable of increasing the farm income.

The system approach came into prominence in World War-II. The system approach can be identified as a scientific method of problem solving, decision making and planning. It a procedure for characterizing the nature of system, so that decision-making might be made in a logical coherent fashion, and the performance of a system might be described.

The weak links between research and extension were identified at many places and the issues like financial sustainability, lack of farmers' participation in programme planning were more prominent leading to some kind of changes in the extension system. During the same period the National Agricultural Research System (NARS) under ICAR has been strengthened through two parallel National Agricultural Research Projects (NARP).

The National Agricultural Technology Project (NATP) was to consolidate the earlier investments and was designed to address the specific system constraints, weakness and gaps those remained unaddressed during the previous attempts in rural extension. The attempt to test innovations in technology dissemination gained momentum to bridge the serious gaps in research-extension-farmer linkage. The research in extension revealed that in private sectors, the presence of a number of organizations providing extension services came up rapidly (Sulaiman and Sadamate, 2000) while, the variation in and among regions in the country and intensification in terms of expenditure, manpower allocation and contract varied widely. The department of agriculture is one the important sources of information for farmers though their role in delivering information on non-food grain crop is limited. The farmers' dependence on other farmers and input dealers as source of information continued to be high reflecting the limited reach of department of agriculture. The organizations like farmers association, producers' cooperatives used to provide a large number of services including extension to farmers but their existence was limited to few commercial crops only. The role of SAU, ICAR, NGOs throughout the country were observed to be location specific with limited activities and funding bound operations.

Looking into the advances and limitation of previous extension systems and making a critical analysis of the system constraints like multiplicity in technology transfer system; narrow focus of the agricultural extension system on farmers focus and feedback; inadequate technical capacity within the extension system; need for intensifying farmers training; weak research-extension linkage; poor communication capacity and inadequate operating resources, "Innovation in Technology Dissemination" as an component of NATP with the support of the world bank came to surface with greater promises to serve the farmers.

The major focus of ITD component of NATP was to initiate and strengthen the process of decentralization, bottom up planning, R-F linkage with a view to improve farmers feedback in the technology development and dissemination process. The component envisaged setting up of institutional innovations at different levels of operation to provide platform to the change.

The Agricultural Technology Management Agency (ATMA) as an institution at district level took birth to reduce the intensity of constraints in blocking agricultural growth. On a pilot basis, ATMAs were created as an autonomous body in 28 districts covering seven states of the country with a view to increase the quality and models of technology available and overcome the limitations being posed over years by the previous forms of technology dissemination. Newer strategy for disseminating newer technologies available with the NARS took reality using the platform of ATMA.

CHAPTER

2

Existing Agricultural Research and Extension Systems

Agricultural research and extension functions are generally organized under the Ministry of Agriculture. However, within the ministry there are separate institutions or departments for performing these functions. These institutions or departments may have different organizational structures and operational procedures. Universities and national research institutes are generally research centres, while the agriculture department performs the extension function.

In this conventional system, most emphasis is laid on breeding, testing and distributing activities. A top-down system is followed in generation and technology transfer, where researchers are expected to come up with better varieties and hand-over them to extension for demonstrations and diffusion to farmers. In this set-up, each function develops its own programme more or less independently, leading to duplication of programs. This is not only a waste of resources but also creates confusion among producers regarding which organization to approach.

This type of research and extension system is hierarchically structured from national level to field level. Internal communications from upper to lower levels within the organization may be easy, but communication between organizations takes a circuitous route and hence is often ineffective. Coordination at lower levels is possible only with specific directives from higher levels.

This model has been successful in meeting the demands of resource-rich farmers and producers of high value commodities, as they are able to communicate their needs to researchers. However, small-scale, resource-poor farmers, especially in less productive and heterogeneous agro-ecological areas, have been left out of this model. There is no feedback from these farmers to agricultural departments and thence to research centres.

Such research extension systems are found most commonly in developing countries. In developed countries, such as the USA, the extension role is also performed by research institutions (Cooperative Extension Services) which facilitates a better integration of research and extension functions. In addition, in such countries the private sector takes an active interest in technology generation and supply.

India has one of the largest agricultural research systems in the world. Currently, the public research system in India is led by the ICAR, which has 4 multidisciplinary national institutes, 64 Central research institutes, 15 National Research Centres (NRCs), 6 Bureaux, 13 Project Directorates, 60 All-India Co-ordinated research projects (AICRPs)/networks and 16 other projects/programmes. In addition, there are 64 State Agricultural Universities (SAUs) and three Central Agricultural University, which operates through 313 research stations. AICRPs are the main link between the ICAR and the SAUs.

The number of centres involved in the AICRPs is about 1,300, of which about 900 are based in agricultural universities and 200 in the ICAR institutes. The ICAR has also Zonal Research Stations (ZRSs) and 200 sub-stations. The National Academy of Agricultural Research Management (NAARM) is another institution under ICAR to conduct research and training in agricultural research management. The ICAR has also established 8 Trainers' Training Centres (TTCs) and 611 Krishi Vigyan Kendras at the district level as innovative institutional models for assessment, refinement and transfer of modern agricultural technologies. In addition, there are 23 general universities under the University Grants Commission (UGC), involved in agricultural research. Several scientific organizations such as the Council of Scientific and Industrial Research (CSIR), Bhabha Atomic Research Centre (BARC), National Remote Sensing Agency (NRSA), Ministries and government departments such as Ministry of Commerce, Department of Science and Technology, Department of Biotechnology, Department of Ocean Development, and more than 100 private and voluntary organizations and more than 105 scientific societies are involved in the agricultural R & E and 3 form the part of the national agriculture research system of India (Vision 2020, ICAR). This extensive agriculture research infrastructure not only conducts agriculture research but are also responsible for educating and providing extension services to the farmers. In the following section, the authors highlighted the agriculture extension system in India with focus on major players providing agriculture extension services in the country.

CHANGING AGRICULTURAL DEVELOPMENT GOALS VIS-À-VIS EXTENSION OBJECTIVES

This section begins with an analysis of three major national agricultural development goals and the role that agricultural extension and advisory systems can play in helping to achieve these different goals. For example, after some Asian nations achieved national food security during the 1980s and 1990s, they began refocusing extension's attention on increasing the production and marketing of high-value crops and products (e.g. China). At the same time, many nations, particularly those in Sub-Saharan Africa, are still not food secure, and this situation may worsen due to high fertilizer costs and the increased use of staple food crops for biofuels within the global food system. In addition, natural resources in many countries are being over utilized, owing to a combination of continuing population growth, increasing demand for agricultural products, and poor farming methods. Therefore, most nations need to help and encourage farmers learn how to integrate sustainable natural resource management practices into their farming systems. Finally, there is growing concern about the potential long-term impact of climate change on agricultural production in many countries, especially those in Sub-Saharan Africa. In this section, we will begin by considering each of these major agricultural development goals.

Achieving National Food Security

A central goal of many countries, especially during the second half of the twentieth century, was to achieve national food security so that urban and rural populations would have adequate food supplies. Increasing the production of basic food crops was the primary focus for achieving national food security during this period, and technology transfer was the primary extension approach used to improve the yields of these staple food crops. Depending on the geographic location of the country, these crops generally included the major cereal crops (e.g. rice, wheat, and maize), roots and tubers (e.g. yams and cassava), and major grain legume crops (e.g. beans and pulse crops), as well as oil seeds. As Green Revolution technologies became available during the late 1960s, many extension systems had a positive impact on increasing agricultural productivity through the transfer of new technologies to all groups of farmers. However, extreme poverty (i.e. less than $1 a day per capita income) remains the central factor affecting household food security (FAO 2006a) and the livelihoods of over 900 million undernourished people worldwide.

Improving Rural Livelihoods

Improving rural livelihoods is now a stated goal among many developing countries. In most cases, achieving this goal involves increasing farm household income, which can both improve household food security and nutrition as well as increase access to health services and education for rural children. However, to achieve this goal, most agricultural extension systems will have to change their strategy, approach, and management structure, as well as upgrade the skills and competencies of their extension staff. Specifically, extension systems will need to begin organizing and training the rural poor so they can successfully pursue new crop, livestock, fisheries, and/or other enterprises that are suitable for local resources, conditions, and market opportunities. In most cases, this will require transforming the traditional top down, technology-driven extension model to a more decentralized, farmer-led, and market-driven extension system. For example, as rapid economic development occurred in many transforming economies, such as China and India, the overall demand for food products began to change, including increased demand for high-value crops such as fruits and vegetables, as well as livestock, fisheries, and other value-added products. Since the economic reforms were first introduced in China during 1979–2007, fruit and vegetable production in China has grown at an annual rate of about 26% a year, and meat products have increased about 20% a year. No other country in the world has ever experienced this level of growth. Much of this growth is due to the size (over 1 million trained extension workers) and strategy of the Chinese extension system (decentralized and more market driven). In summary, to meet the changing demand for both staple and high-value food products, extension systems must broaden their focus and teach new technical, management, and marketing skills. This change in strategy will enable small-scale men and women farmers to take advantage of new market opportunities and the changing worldwide demand for both staple and high-value food products.

Improving Natural Resource Management

The natural resources of many countries are under increasing stress, and many nations are becoming more concerned about achieving environmental sustainability through efficient use of land and water resources. Given continuing population increases and the pressures of economic development, national governments must carefully monitor their natural resources and take the necessary actions to maintain them. For example, the agricultural sector typically uses up to 70% of a nation's water resources, but with increasing urbanization and industrial development the water resources of many nations are being over utilized, with long-term negative consequences. Therefore, farmers must learn how and be convinced to use more water-efficient technologies and/or to shift to more water-efficient crops. Some technologies, such as water harvesting, require more labour inputs, whereas most irrigation technologies (e.g. drip irrigation) require substantial capital investments and higher operating costs.

Other technologies, such as integrated pest management (IPM), can help maintain natural resources while reducing production costs. However, disseminating many of these technologies or production practices such as IPM will require a substantial increase in non-formal education services, such as those delivered through Farmer Field Schools (FFS). Most national extension systems do not have sufficient numbers of well-trained extension staff or the financial resources (without continuing donor support) to conduct 8–12 weekly FFS during a single season for 15–20 farmers (i.e. FFS is a labour-intensive methodology). Finally, the lack of an adequate transportation infrastructure plus rising energy costs have increased the cost of fertilizers in many countries. These factors, along with low staple food prices, has made it difficult for small scale farmers in many countries to maintain or increase their productivity levels while also maintaining their macro- and micro-nutrient soil fertility levels.

TYPES OF AGRICULTURE AND RESEARCH-EXTENSION LINKAGE

The nature of research-extension linkage problems vary with the agroclimatic and socioeconomic characteristics of the local agriculture (Chambers, 1988). Based on the above characteristics, Chambers classified agriculture into three categories.

The first category is characterized by high-input, high-yielding production systems, and found primarily in developed countries and in a few areas in developing countries. The problems of farmers in this system are overproduction and escalating costs. Here the research-extension linkage is well established.

The second category consists of areas benefitting from the green revolution, where applied research has made a dramatic impact on foodgrain yields. This type of agriculture is found mainly in high-capacity areas of the tropical countries, and particularly where irrigation is available. In these areas, a reasonably strong linkage between research and extension exists, and the need for further strengthening the linkage is moderate.

In the third category, which is characterized by poor and diverse resources endowments and ecological conditions, the most urgent need is to strengthen the research-extension linkage. These areas are characterized by low and uncertain rainfall, lack of irrigation, and poor infrastructure. Yields are low and uncertain, and there is a general degradation of resources. However, the potential for increasing agricultural production is high and unexploited. The challenge in these areas is to develop sustainable technology for the heterogeneous agroecological and socioeconomic conditions.

Though attention has been given to the third category of agriculture, most prevailing research and extension systems have been organized and managed to meet the technology demand of the first two types, as they generate urgently needed foreign exchange and produce foodgrains. The technology needs of the third category of farmers are mainly expressed through government policy, as the farmers are unorganized in such areas.

Linking Decentralization of Agricultural Extension Systems to Sustainable Agricultural Development

Decentralization is defined as the transfer of effective control by central managements to regional and provincial managements or other field level offices. In addition, this strategy may include the participatory involvement of farmers in the managerial processes for agricultural development (Rivera et al. 1997). District extension director received and followed instructions from the senior management of the agricultural extension with limited involvement of subordinate staff. The staff are involved in the development of the case organization's annual extension plan and each staff member is responsible in consultation with his supervisor, for the development of his own annual work plan and training program. Two field staff representatives are also included in a management team comprising the director and assistant, the supervisors and a support staff representative (Okorley et al., 2009). Prior to decentralization, the management of the case organization was top-down-the decentralization is an example of promoting the participation of lower-levels of agricultural extension management in decision-making and budgeting. And extension, participatory and demanded services are examples of the effort to integrate producers into agricultural processes (Niamh Dennehy et al., 2000), as this allows much greater transparency of decision-making because the field staff representatives are involved in the actual decision-making (Okorley et al., 2009). Decentralization also encourages more contact and open communication to build respect and trust among the staff, gives a level of flexibility to field staff to design their location-specific extension activities with farmers. It encourages team work amongst the staff, and has opened itself up to increased scrutiny and input from farmers and other stakeholders through greater interaction with them (Okorley et al., 2009). This is undertaken to improve the field staff's knowledge of farmer practices and the reasons behind these practices to foster this learning culture. The case organization provides a range of mechanisms through which staff can learn informally, such provides learning materials that the staff can access for self-directed learning. It creates an open environment in which staff feel comfortable in sharing information, as such this provides support to the field staff in decision-making, and encourages teamwork among the staff; and ensures that the staff are informed in a timely fashion about policies and other relevant issues (Okorley et al., 2009). The needed reforms include decentralization of responsibility, delegation of authority to district managers and teams, autonomy in routine decision-making, and a separate budget for operational expenditure. To adopt new technologies, solve problems, and increase income from agriculture, must have to reorganize its structure and functions by embracing wider expertise, decentralizing management, and nurturing a culture of organizational learning (Van den Ban and Wageningen, 2003). It should take into consideration the diversity of organizations that are providing different extension services and the potential for improving the relationships among them. While extension managers and policy makers need to explore these options for providing better extension services to farmers to meet the emerging challenges (van den Ban and Wageningen, 2003). The technologies developed were often inappropriate for small-scale farmers, as the conditions on-farm, including the farmers' own management type and priorities, were not adequately considered (Davis, 2008). Understanding of human resource capacity building is a key factor of success for decentralized public agricultural extension and other institutions such as research institutes, universities and other government organizations to facilitate training. This proximity to major research institutions provides it with an advantage in relation to accessing expertise for training. The critical feature of field staff training at the case organization is the involvement of farmers in the training process, a practice they call "joint training" exercise (Okorley et al., 2009). Institutional reform has resulted in a variety of institutions being engaged in the transfer and exchange of agricultural information; as well as institutional reform through privatizing schemes such as contracting with the private sector and the establishment of partnerships in the provision of agricultural extension services (Kim et al., 2009). There is no way the private sector organizations can effectively provide extension services without the assistance of the state and also from agricultural development organizations, because they already have well-trained personnel and infrastructure in place (Kristin Davis and Place, 2003). Extension and research staff will be accountable to farmer clients through the participation of farmer organizations and emerging agricultural structures in decision-making processes, and supported to ensure that they have a say in formulating policies that affect them (Al-Rimawi and Al-Karablieh, 2002). Other intervention measures include providing effective information dissemination to farmers, improvement in technology delivery mechanisms and increasing outreach such as making technology component farmer specific. Others are decentralization of agricultural technology delivery institutions, enhancing farmer's managerial ability especially through farmers' organizations and educational institutions and reforming agricultural markets to stabilize income of farmers (Chukwuone et al., 2006). Consequently, an increase in the quality and quantity of adult and continuing education programme is a priority and educational institutions are charged with the task of designing programme curricula to achieve these policy aims. Higher education today operates in a new era, an era that is much more conscious of the market place (Angstreich and Zinnah, 2007). Towards this end, it is necessary to review the potential of developing measures for the greater organizations when it comes to the agricultural extension organizations, design agricultural extension organizations in the regions centered on the key products, and various alternatives (Kim et al., 2009).

The Move towards Pluralism for Sustainable Agricultural Development

Agricultural extension managements can establish different collaborative working relationships with agricultural development organizations based on trust and mutual respect, to obtain access to resources for extension delivery farmers and staff training (Ernest et al., 2010). The main challenge in installing a proper pluralistic agricultural extension mechanism is the effective coordination among various organizations, especially in matters of development when competent nonpublic institutions are present in the country (Rivera and Alex, 2004). The modality of using more than one organization, whether public or private, for delivering extension services is to help in achieving the desired goals (Rivera and Alex, 2004). In addition, agricultural research institutes, agricultural universities and farmers' associations, participate in the delivery of extension services. Here, agricultural extension refers to the cultivation of farmers' organizations that aim to increase agricultural productivity and to improve the everyday life of farmers (Ban and Hawkins, 1988). The agricultural technology distribution is a model which shows the relationship among agricultural research, agricultural extension and farmers (Ban and Hawkins, 1988). Based on the agricultural technology distribution, agricultural extension process is a scientific knowledge the results of agricultural research to the techniques and transmits the techniques to the farmers to help them adopt the techniques and increase production by using them (Kim et al., 2009). Agricultural research and technology identification are often relevant to all public and private extension service providers. Here, most extension services oversight is an inherent aspect of the public sector's responsibilities for policy formulation, and design of reforms to promote pluralistic extension institutional arrangements (Rivera and Alex, 2004). The obvious rationale is the pooling of all available resources in order to alleviate pressure from low budgets and staff in the ministries of agriculture, as well as to let the farmers benefit from a variety of sources (Rivera and Alex, 2004). But pluralistic extension also requires of emphasizing multiple and diverse partnership between public and private sectors including partnership with farmer organizations and private venture companies to facilitate the common concepts, language, methods and skills needed to integrate the diversity that arises from institutional pluralism (Rivera and Alex, 2004). As farmer organizations mature, they may become increasingly oriented toward providing specific services for their members (Burton E. Swanson and Rajalahti, 2010). Farmers' associations have long played an important role in providing advice on production technologies, and putting pressure on research and extension organizations to work in a more demand-driven and client-oriented way (Van den Ban, 2000). The success of agricultural extension and development projects often depends on local participation, because this enable them to work as partners in planning and implementing agricultural and development programs. Extensionists should develop programs that facilitate the use of new technology to optimize this process and encourage farmers to use new technologies along with extension and other organizations (López and Bruening, 2002). Institutional reorientation can be achieved by strengthening farmer organizations to have a decisive role in determining extension agendas, programs, and services through contracting, decentralization, and support to local innovation (Rivera and Alex, 2004). The proposed organizational linkage structure is intended to promote cooperation and coordination among development organizations through involvement of the farmers, thereby providing a more structured and permanent basis for interaction between the organizations involved (Duvel, 1996). The organizational linkage must have maintained good collaboration among agricultural extension and farmers' organizations, and provide support in training the farmers and in the implementation of various agricultural projects (Kim et al., 2009). This structure is not intended to be an alternative institutional framework for agricultural extension that focus on the implementation of development programs or projects; which usually consists of top management, middle management, support staff and field or site staff. It is visualized as a linkage structure or system with the purpose of linking development organizations with the community in an effective partnership (Duvel, 1996). Another trend is the formation of agricultural organizations (which are less bureaucratic, more flexible and with wider expertise) to implement special programs related to agricultural development (van den Ban and Wageningen, 2003). Farmer associations and producer cooperatives are also presently involved in agricultural extension services, because they are the most effective in reaching farmers producing these crops and commodities (van den Ban and Wageningen, 2003). Farmer organizations should therefore be a high priority for public sector extension, because farmers need a wide range of services related to technology (production and processing), quality, access to markets, price information, and business development, and improve the ability of farmers to collectively find solutions to their problems (van den Ban and Wageningen, 2003). Economic and social issues require the allocation of appropriate resources for this work; this entails a commitment to develop proficiency local groups to participate in these processes (Niamh Dennehy et al., 2000). Farmers belonging to farmers organizations are more aware of the constraints they were facing to improve their production than nonmembers. This may be due to the fact that most extension programs were intended for farmers' organizations instead of individual farmers (Owona Ndongo et al., 2010). The involvement of public organizations in institutional research and extension activities can lead these institutions to establish complementary relationships with such organizations as the Agricultural Research Institutes, the Ministry of Agriculture, and similar agricultural development organizations (Teffera Betru and Hamdar, 1997). As the cost of research is high, the public system is more technically and logistically equipped to undertake research activities, and the firms have direct interest to cooperate with the public research in undertaking experimental works (Al-Rimawi and AlKarablieh, 2002). The provision of extension assistance to farmers previously supported by participating organizations and the development of seed supply networks that are accessible and affordable to subsistence farmers represent two tangible areas where linkages between public and private extension activities could provide important benefits (Rodney Reynar et al., 1996). Development programs worldwide have recognized that local participation is the key to sustainable transfer and long-term adoption of new technologies and approaches. Interactive participation is the approach that facilitates this kind of learning environment. Teaching has long been the normal mode of educational programs and institutions where agricultural extension skills work (Toness, 2001). The capacity building component was designed around three objectives: To develop competency-based curricula in participating universities that better match agricultural sector workforce needs. To develop new and updated courses, and improve instruction; and to develop internship programs to provide real-life experiences working with farmers, exporters and other agribusiness firms for college graduates (Barrick et al., 2009). In order to move from a teaching paradigm towards a learning paradigm, highly participatory interactions and knowledge sharing among all sectors is critical for extension institutions both in applied extension programs and at teaching institutions. Emphasizing the strengths of both public and private extension initiatives may begin to fully address the needs of subsistence farmers (Toness, 2001). A case is made for the organizations involved to continue to cross the institutional divides so that the long-term sustainability and development of small-scale farming communities is ensured. Conventional station-based approaches to agricultural research, technology development, and extension have failed to achieve the expected results in the small-scale farming sector of the developing world (Davis, 2008).

Reforming the Agricultural Extension System in India

DEFINITION AND CONCEPTS OF REFORMS

Reform means the improvement or amendment of what is wrong, corrupt, unsatisfactory, etc. (Wikipedia). The use of the word in this way emerges in the late 1700s and is believed to originate from Christopher Wyvill's Association movement which identified "Parliamentary Reform" as its primary aim.

As stated in Wikipedia, reform is generally distinguished from revolution. The latter means basic or radical change; whereas reform may be no more than fine tuning, or at most redressing serious wrongs without altering the fundamentals of system. Reform seeks to improve the system as it stands, never to overthrow it wholesale. Radicals on the other hand, seek to improve the system, but try to overthrow whether it be the government or a group of people themselves.

According to Merriam Webster Dictionary, reform means to improve (someone or something) by removing or correcting faults, problems, etc. or to improve your own behavior or habits.

As stated in Cambridge Dictionaries online, reform is to make an improvement, especially by changing a person's behavior or the structure of something.

According to Oxford Dictionaries, "reform" is defined as to "make changes in (something, especially an institution or practice) in order to improve it".

Agricultural extension in India has undergone several changes since independence. Still, a large number of small holder farmers and other vulnerable groups remain unreached by the public extension system. A number of organizational performance issues hinder the effectiveness and efficiency of public extension system. These include inadequate staff numbers, low partnerships and continued top-down linear focus to extension.

The Indian public agricultural extension system is one of the largest knowledge and information dissemination institutions in the world. Several efforts have been made in the public sector over the past one decade to initiate various reform measures and operational models to improve the organizational performance of this system. Yet, the challenge of enhancing relevance, efficiency and effectiveness of the public sector agricultural extension system in meeting its organizational goals and objectives remains unresolved (WGAE, 2007; Raabe, 2008; Glendenning et al., 2010; Desai et al., 2011). India is not alone in the world in reforming its extension and research systems.

There are many countries where extension and advisory services reforms are occurring globally (Swanson and Rajalahti, 2010; World Bank, 2012).

A considerable variety of public sector reform strategies have emerged, and can be categorized in different ways. Porter (2001) in his study organized reform under two main headings: market reforms and non-market reforms. According to this distinction, market reform strategies: revision of public sector extension system, pluralism, cost recovery and total privatization. Non-market reforms comprise two main reform strategies: (a) decentralization, transferring central government authority to lower tiers of government and (b) subsidiarity, transferring or delegating responsibility, sometimes by abolishing authority over extension, to the lowest level of society as is practical and consistent with the overall public goods (Porter, 2001).

Figure 3.1 provides a more dynamic view of institutional extension reforms, and further illustrates that an extensive menu of options exists for governments to consider in any agricultural and rural extension reform. It also suggests that there is no single solution to the question of reform. Indeed, several areas of reform might be combined to formulate a country's policy, depending upon that country's situation and the way the government views its needs (Rivera, 2001).

In order to realize agricultural potential and to increase agricultural yields, India's extension system has experienced major conceptual, structural, and institutional changes since the late 1990s. This book reviews existing reform programs and strategies currently existing in agricultural extension in India. It distinguishes strategies that have been employed to strengthen both the supply and demand sides of service provision in the area of agricultural extension, and it reviews the effects of the demand- and supply-side strategies on the access to and the quality of agricultural extension services. The ultimate objectives are: (1) to gain a view on what works where and why in improving the effectiveness of agricultural extension in a decentralized environment; (2) to identify measures that strengthen and improve agricultural extension service provision; and (3) to reveal existing knowledge gaps. Although the range of extension reform approaches is wide, this book shows that an answer to the question of what

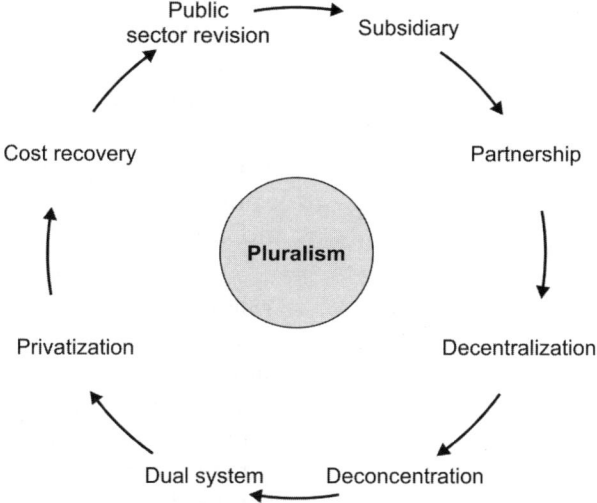

Fig. 3.1: A dynamic view of extension institutional reform. Source: Rivera (2001)

works where and why is complicated by the absence of sound and comprehensive qualitative and quantitative impact and evaluation assessment studies. Even evidence from the National Agricultural Technology Project and the Diversified Agricultural Support Project of the World Bank, the women empowerment programs of the Danish International Development Agency, the Andhra Pradesh Tribal Development Project, and the e-Choupal program of the Indian Tobacco Company is subject to methodological and identification problems. Policy reforms of agricultural extension in India emphasizes (i) the importance of implementing both decentralized, participatory, adaptive, and pluralistic demand- and supply-side extension approaches; (ii) involving the public, private, and third (civil society) sectors in extension service provision and funding; and (iii) strengthening the capacity of and the collaboration between farmers, researchers, and extension workers are necessarily tentative and require further quantification. This chapter seeks to inform policymakers and providers of extension services from all sectors about the need to make performance assessments and impact evaluations inherent components of any extension program so as to increase the effectiveness of extension service reforms.

RELATIONSHIP BETWEEN NATIONAL AGRICULTURAL DEVELOPMENT GOALS AND DIFFERENT AGRICULTURAL EXTENSION OBJECTIVES AND FUNCTIONS

As governments consider how to strengthen their extension systems to achieve their national agricultural development objectives, they need to consider how these different extension functions relate directly to these overall national goals, as illustrated in Fig. 3.2—Each of these key functions is described in more detail in this section.

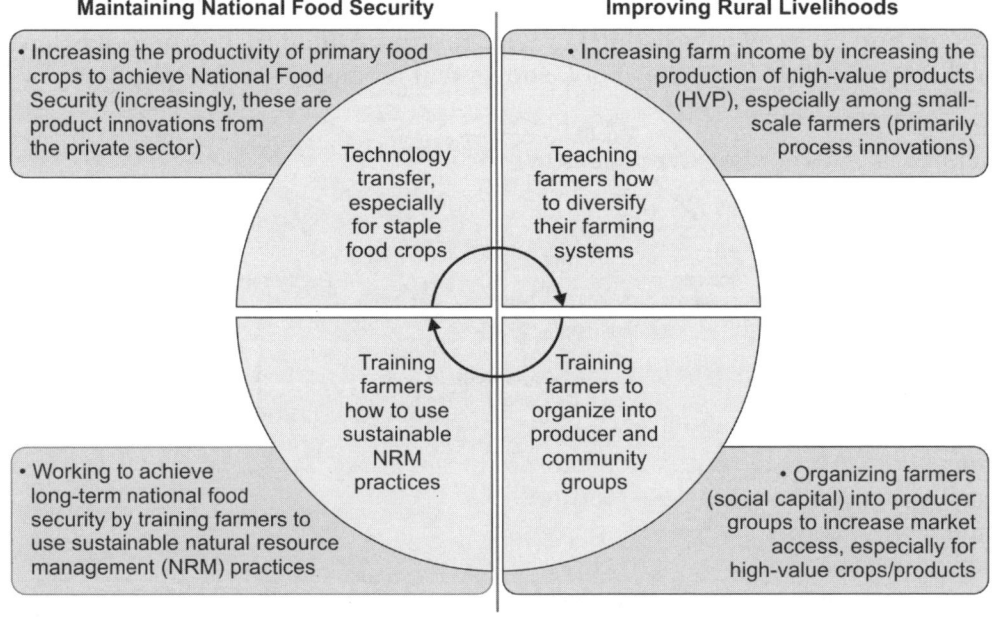

Fig. 3.2: Key extension service function vis-à-vis National Agricultural Development Goal. *Source:* Swanson and Rajalahti (2010)

Objective 1: Transferring New Agricultural Technologies to Achieve National Food Security

During the second half of the twentieth century, most national extension systems primarily focused on transferring agricultural technologies that would increase the productivity of major crop and livestock production systems in achieving national food security. This primary extension objective was greatly reinforced and enhanced during the Green Revolution, when improved technologies, especially for wheat and rice, were transferred to the many farmers who benefitted, particularly in Asia.

When considering technology transfer as an extension strategy, it is useful to briefly review the basic concepts outlined in Everett Rogers's classic 1962 book, *Diffusion of Innovations,* which is currently available in its fifth and final edition (2003). As he pointed out, the first adopters of new, research-driven innovations (i.e. new technologies) are generally the more progressive, commercial farmers who were classified as *innovators* (about 2 to 3%) or *early adopters* (about 13 to 14%). Therefore, during the twentieth century, especially in industrially developed countries, it was commonly accepted that in pursuing this technology-transfer extension approach, it would be the larger, better-educated farmers who would be among the first group to adopt these innovations (i.e. technologies). Medium-scale commercial farmers (about 34%) fell into the *early majority* category, while smaller-scale and subsistence farmers generally fell into the categories of *late majority* (about 34%) or *late adopters* ("laggards," about 16%). In the dissemination of Green Revolution technologies, this same adoption pattern occurred, but the process took place much more rapidly, especially the spread of new, high-yielding varieties of wheat and rice, together with the necessary production practices (e.g. plant population and tillage practices) and inputs (e.g. fertilizers and agrochemicals).

Another important concept to keep in mind is that during the twentieth century, especially in Europe and North America, most extension professionals accepted the fact that most small-scale farmers would not be competitive in a dynamic agricultural economy. It was generally understood that the majority of these smaller, low-resource farmers (or their children) would eventually leave farming as large-scale commercial farmers captured more of the profits from new technologies and as high-resource, commercial farmers expanded their farming operations.

The resulting rural–urban migration was not considered a serious problem in most industrialized countries, given the concurrent rapid growth of the industrial and service sectors that, in effect, pulled many rural people into urban jobs. However, this "push–pull" phenomenon is not occurring so rapidly in most developing countries, especially in Sub-Saharan Africa. Therefore, the immediate goal is to pursue a more balanced extension strategy, including market-driven innovations, which will help increase farm incomes and thereby improve rural livelihoods. Under this more balanced extension strategy, as illustrated in Fig. 3.2, the goal will be to help small-scale farm households, especially among the rural poor, improve their livelihoods by increasing their farm income, achieving household food security, organizing into producer groups (i.e. empowerment), and increasing their access to health services and education for their children.

In considering this broader goal of improving rural livelihoods, public extension should no longer give priority to the traditional "diffusion of innovation" approach of transferring improved technologies that will provide the greatest economic benefit to larger, commercial farmers. This task is rapidly being taken up by the private sector through input supply dealers. In addition, the costs of this new and expanding source of "technical advisory services" will be progressively shifted to the farmers themselves, especially those larger commercial farmers who use more inputs. Instead, public extension should give more attention to a broader extension strategy that includes more attention to changing markets for high-value crops and products, organizing farmers into producer groups to supply these markets, and using more sustainable natural resource management practices.

In addition, there can and/or should be close cooperation between public extension (especially subject-matter specialists) and private input suppliers, because most local input dealers, especially at the district and sub district levels, do not have technically competent sales personnel who can give correct technical advice to farmers. Therefore, rather than public research and extension personnel viewing the private sector as competitors, they should develop public–private partnerships, especially with interested input supply dealers.

Objective 2: Increasing Farm Income to Improve Rural Livelihoods

Currently, many nations and some agricultural extension systems are shifting their attention to the broader goal of improving rural livelihoods. To achieve this goal, national extension systems will need to enhance the technical, management, and marketing skills (i.e. human resource development) of all farmers, but especially small-scale men and women farmers, as well as the landless, indigenous populations, rural young people, and other vulnerable groups. The task is to help these different households select and successfully produce an appropriate mix of crop, livestock, and/or other enterprises that is most suitable for their location (access to markets); agro-ecological conditions; and land, labor, and water resources.

In considering how best to implement the extension objective of improving rural livelihoods, it is necessary to differentiate among types of farm households (i.e. subsistence; small-scale; medium-scale; and larger, more commercial farmers) and to consider traditional differences between men and women farmers, as well as rural young people. For example, many small-scale subsistence farmers, particularly women farmers, usually lack basic education; therefore, their needs will differ substantially from the skills and knowledge needed by medium-scale and, especially, commercial farmers. In addition, the role of women farmers within households differs considerably across different cultures, agro-ecological zones, and farming systems; therefore, the needs and opportunities for each category of farmers must be carefully examined. For instance, different examples will be provided about how rural women learned how to use common property resources (village ponds and nearby forests) to begin producing different high-value products, such as freshwater fish and silk cocoons. In addition, there are many other successful examples of how landless household members, including rural women, learned how to produce and market other high-value products, such as backyard broilers, eggs, mushrooms, honey, vermicompost, and so forth. Finally, rural young people have largely been ignored by most national extension systems in the past, but many of these young people will be the future men and women farmers in most rural communities.

Objective 3: Organizing or Empowering Farmers by Building Social Capital within Rural Communities

In most developing countries, public extension systems have not traditionally been very interested in organizing men and women farmers, including rural youth, due to demands on the extension system itself and/or the Ministry of Agriculture (i.e., more inputs, credit, etc.). In addition, because extension's primary focus has been on technology transfer for the major food crops, building social capital did not play an instrumental role in this earlier agricultural development strategy. However, in helping improve rural livelihoods, it will be important, if not essential, to organize farmers, including women farmers, into different types of producer groups and then help to link these groups to markets for appropriate high-value crops and products in addition to other information and organizations, such as research. Failure to do so may result in other value chain actors continuing to capture the majority of the profit from these high-value enterprises, while farmers continue carrying the risk of producing high-value, perishable products.

Furthermore, as noted above, organizing rural youth groups can be an effective, long-term strategy for building both human and social capital within rural communities. This approach continues to be a top priority in a few public extension systems worldwide, such as Costa Rica, Nigeria, Tanzania, Thailand, and the United States.

Objective 4: Training Farmers to Use Sustainable Natural Resource Management Practices

During the past 20 years, worldwide expansion of arable cropland has diminished considerably. At the same time, the world's population is expected to reach nearly 9 billion by 2050; therefore, global food production will need to double during this period, if world hunger is to be reduced. In addition, as mentioned earlier, there is an on-going shift from fewer cereals to increased meat, milk, fish, vegetable, and fruit consumption in many Asian and Latin American, and more recently in some Sub-Saharan African countries. These changes, combined with the overconsumption and/or waste of food products by affluent consumers, may result in total food demand increasing by upward of 2.5 times over current production levels. It should be noted that the global production of cereal crops, on a per capita basis, peaked during the 1980s and has been slowly decreasing since then, despite annual increases in average yields (UNEP 2007a, p. 110).

Moreover, the world's natural resources for food production are under considerable pressure. For example, soil nutrient depletion is occurring in many tropical and subtropical countries, and water scarcity is already becoming more acute in many regions, especially where farming takes the lion's share of water being withdrawn from streams and underground aquifers. Other claims on scarce water resources are growing rapidly, particularly from industrial development and the growth in urban populations worldwide. Likewise, desertification, land degradation, and the excessive use and pollution of underground aquifers continue to occur in many countries.

It is clear that there is an urgent need for public extension systems in most countries to give higher priority and to allocate more resources to educating farmers how to use low-cost, sustainable natural resource management practices. If not done, there will be serious, long-term consequences for many countries as these natural resource management problems become more acute and as total food demand increases. Farmers must first understand these long-term consequences and then learn how best to address these NRM problems. However, many farmers will have neither the incentives nor the resources to adopt sustainable NRM practices unless they first learn how to diversify and/ or intensify their farming systems as a means of increasing farm income. For example, small-scale farmers can be encouraged to adopt drip irrigation technology, if they are able to produce and market high-value horticultural crops. Likewise, farmers in some countries are moving to zero grazing livestock systems, so they can both increase the fattening rate and then utilize the manure to produce *organic* farm products.

STRENGTHS OF INDIAN EXTENSION SYSTEM

India is in process of transforming its agricultural extension and technology transfer systems to become more demand-driven and responsive to farmers' needs. There is need to develop skill and knowledge on scientific agriculture. Its wide extension system could be visualized through these facts (1) India has second largest extension system in the world in terms of professional and technical staff. More than 90,000 technical personnel constitute its extension system (Brewer, 2000). Hence, needs to utilize these large human resources in the effective transfer of technology process and (2) there are 100 million farm families supported by the large agricultural extension services, which is financed by state governments. Since independence, it has used different extension approaches with mixed results supported by over 90,000 staff members (Swanson and Mathur, 2003).

WEAKNESSES/CONSTRAINTS OF INDIAN EXTENSION SYSTEM

Existing weaknesses/constraints in Indian agricultural extension system are mentioned as the problems and constraints of extension system as identified by Singh *et al.* (2006) are: (i) Top-down approach (ii) Being commodities and supply-driven specific (iii) Declining farm income (iv) Lack of farming system approach (v) Accountable to government than farmers (vi) Weakening research-extension linkages, and (vii) Little focus on empowering farmers.

Swanson and Mathur (2003) reviewed agricultural extension system constraints as; (i) Multiplicity of public extension systems (ii) Narrow focus of agricultural extension system (iii) Co-mingling of government schemes and extension activities, (iv) Lack of farmers involvement in extension program planning (v) Supply rather than market-driven extension (vi) Lack of transparency and accountability (vii) Inadequate technical capacity (viii) Lack of local capacity to validate and refine technologies (ix) Lack of emphasis on farmers training (x) Weak research-extension linkage (xi) Weak public sector linkages with private sector firms (xii) Inadequate communication capacity (xiii) Inadequate operating resources and financial sustainability. (xiv) Since T & V system ended, there has been little donor support for extension, and reliance almost solely on state government funding. Extension system of 1990s has been described as weak, ineffective and inefficient (Raabe, 2008 and Suman, 2014). Extension services are characterized by biases that result in tending to neglect poor farmers, particularly women. There has been a wide range of chronic problems in public sector (Bharati et al., 2014). (xv) High staff vacancy rates, low social status, low rank in the administrative system, lack of operational funds for effective field work and high turnover were reported by Birner and Anderson (2007).

Major constraints emphasized in 11th Five Year Plan recommendations were: (i) Lack of convergence in operationalization of extension reforms (ii) Lack of provision for dedicated manpower at various levels (iii) Inadequacy of funds, (iv) Lack of infrastructural support below district level, and (v) Inadequate support for promotion of farmers' organizations and their federation.

OPPORTUNITIES

Public sector extension in both developed and developing countries is undergoing major reforms. Agricultural extension continues to be in transition as governments and international agencies are advancing structural, financial and managerial reforms to improve the pluralistic extension system. Decentralization, pluralism, cost sharing, cost recovery, participation of stakeholders are some of the elements in extension's current transition. Views on extension have changed from an agency of technology dissemination with emphasis on agricultural production to helping farmers organize themselves, linking of farmers to markets (Swanson, 2006) and providing environmental and health information services. The recent reform-oriented initiatives have been directed towards creating a demand-driven, broad-based and holistic agricultural extension system (Planning Commission, 2008). This has involved the design and introduction of a multitude of integrated measures that—on the demand side-enable service users to voice their needs and hold service providers accountable, and on the supply side influences the capacity of service providers to respond to the needs of the extension service users (i.e. the farmers).

CHALLENGES/THREATS

In current scenario, where a numbers of stakeholders are involving in agricultural extension, hence, opportunity to reach a greater number of farmers is increasing. In this context, private sector is incorporating extension services within existing service provisions and experimenting with ICT. But inherent challenges each sector faces in reaching different farmers means that partnership and coordination between sectors will best serve the interests of farmers. Hence, addressing of current challenges is necessary.

EMERGENCE OF PLURALISTIC EXTENSION SYSTEMS IN INDIA

Several institutional innovations have come up in response to the weaknesses in these public research and extension systems that have given enough indications of the emergence of an agricultural innovation system in India. This has resulted in the blurring of the clearly demarcated institutional boundaries between research, extension, farmers, farmers' groups, NGOs and private enterprises. Figure 3.3 envisages pluralistic extension system should play the role of facilitating the access to and transfer of knowledge among the different entities involved in the innovation systems and creates competent institutional modes to improve the overall performance of the innovation system. Inability to play this important role would further marginalize extension efforts. In India, the main agency for agricultural development is Union Ministry of Agriculture at national level and the state Departments of Agriculture. In the first line extension system, the Indian Council for Agricultural Research (ICAR) and the State Agricultural Universities (SAUs) play a major role through organizing demonstrations, training, etc. on a limited scale, but forceful enough to have a catalytic influence on other extension systems and sub-systems. The detail description of various agencies at central, state and district level has been depicted in Fig. 3.3. However, brief description regarding the institutions/agencies involved in pluralistic extension system is presented below.

Central Level

The Union Ministry of Agriculture (www.agricoop.nic.in)—a branch of the Government of India, is the apex body for the formulation and administration of rules and regulations and laws relating to agriculture in India. The Union Ministry of Agriculture comprises Department of Agriculture and Cooperation (DAC), Department of Agricultural Research and Education and Department of Animal Husbandry, Dairying & Fisheries. Secretary, Agriculture & Cooperation is the administrative head of the department and is responsible for formulation and implementation of policies of Agriculture and Cooperation. The DAC is responsible for formulation and implementation of national policies and programmes. The record production of food grains has been achieved through effective transfer of latest technologies and development schemes being implemented by the Department of Agriculture & Cooperation backed by remunerative prices for various crops through enhanced minimum support prices. The institutions engaged in extension of agricultural activities under the public sector have been discussed below.

Department of Agriculture and Cooperation (DAC)

The DAC (http://agricoop.nic.in/add.htm) is committed to the welfare and economic upliftment of the farming community in general. The Department formulates and implements National Policies and Programmes for achieving rapid growth and development through optimum utilization of country's land, water, soil and plant resources. The DAC comprises several technical directorates (also called divisions) and one of them is for agricultural extension. The Directorate of Extension (DoE), headed by a Joint Secretary cum Extension Commissioner, is the nodal extension organ at the national level. The Joint Secretary is assisted by three Joint Commissioners. The directorate provides policy guidelines and operational backstopping to the state level extension organizations. At times, it has directly implemented certain major programs. DoE organises agriculture fairs at national and state level. It also offers model training programmes to develop the skills of the state extension functionaries. It support to the schemes namely, (i) Central sector scheme on extension support to central institutions (ii) Revised ATMA scheme (iii) Mass media support to agricultural extension (iv) Revised schemes of Agri-clinics and agri-business centre. The Directorate of Extension (DoE) (http://vistar.nic.in) was set up under DAC in 1958 in the wake of launching of Community Development Programmes and National Extension Service throughout the country in 1953. Apart from functions of dissemination of specific knowledge to farmers and supervision of the countrywide extension training infrastructure, DoE was also later called upon to implement National Programmes like Intensive Agricultural District Programme (IADP) and Intensive Agricultural Areas Programmes (IAAP). However, since 1974 the emphasis was shifted to Training and Visit system of Extension, which was introduced in 17 major states with the World Bank Assistance. Its role is essentially collaborative, providing guidance and technical support to the Extension Division. The directorate's technical units are extension management, extension training, farm information, and National Gender Resource Center in Agriculture (NGRCA). The Extension Education Institutes (EEIs) were established at 4 locations, i.e. Nilokheri (Haryana in 1958), Rajendranagar, Hyderabad (A.P. in 1962), Anand (Gujarat in 1962) and Jorhat, (Assam in 1987) on regional basis to meet the training requirement to middle level extension functionaries of States and Union Territories as well. National Institute of Agricultural Extension Management (MANAGE) (www.manage.gov.in) located in Hyderabad, Andhra Pradesh (AP)—is an autonomous organization established by the government in 1986 for assisting the central government and the state governments to improve their pluralistic extension systems by bringing positive changes in policies, programs, and personnel skills. Main activities undertaken by the institute are extension capacity building, research, consultancies, education in management, and documentation. This institute offers dozens of training courses advertised well in advance. It also offers two postgraduate diploma programs, one in general management and the other in agricultural extension management. In addition, a one year diploma program in agricultural extension services for input dealers was started in 2004 for imparting formal agricultural education to the dealers. MANAGE is also responsible for implementing the Agri-Clinics and Agri-Business Centers Scheme (ACABC), which aims at providing value-added extension services to the doorsteps of farmers by agricultural professionals. The scheme involves two-month residential training to eligible agricultural professionals, one-year post training in handholding support, start-up loans by banks and subsidy by the National Bank for Agriculture and Rural Development (NABARD). MANAGE enjoys highly qualified and experienced faculty and well equipped modern training infrastructure. Its training programs are open to both public and non-public stakeholders.

Indian Council of Agricultural Research (ICAR)

The ICAR (http://www.icar.org.in/en/)—is the apex body for coordinating, guiding and managing research and education in agriculture including horticulture, fisheries and animal sciences in the entire country. The ICAR has played a pioneering role in ushering Green Revolution and subsequent developments in agriculture in India through its research and technology development. It has played a major role in promoting excellence in higher education in agriculture. It is engaged in cutting edge areas of science and technology development and its scientists are internationally acknowledged in their fields. The Agricultural Extension Division (http://www.icar.org.in/en/agricultural-extension.htm) which is a part of the ICAR is headed by a Deputy Director-General (Agricultural Extension), who is supported by two Assistant Director-Generals. Activities of this Division are technology assessment and demonstrations, training of farmers, training of extension staff, and creation of awareness of improved technologies among farmers. The division performs extension activities through the following institutional mechanism. There are 44 Agricultural Technology Information Centres (ATIC) established under ICAR institutes and SAUs.

There is one Directorate of Research on Women in Agriculture (DRWA) located in Bhubaneswar (Odisha). Extension division monitors the extension activities carried out by KVKs through 8 Zonal Project Directorates (ZPDs) across the country.

Civil Society—Farmers Organizations, Associations and Societies

Civil society organization includes Farmers' Associations, Cooperatives and Societies employed in extension of agricultural activities. In India, these organisations have been fairly active for years aiming self-help for development, specific commodity production, marketing, collective bargaining and many other purposes. In India, self-help groups are playing greater role in transfer of agricultural technologies (Meena et al., 2003; Meena et al., 2008; Meena et al., 2011).

Major emphasis has played on poverty alleviation and rural women empowerment. Farmers' association's examples are: Punjab Young Farmers Association (India); Indian Farmers Association; Turmeric Farmers Association of India; Farmers' Association Pomegranate; Association of Farmer Companies; Organic Farming Association of India (OFAI) and many more. Nearly 580,000 cooperatives are functioning in India in addition to 375,000 agricultural cooperatives with 280 million member farmers. Agricultural cooperatives deal in credit, inputs, marketing, agro-processing and farm extension services. There are fertilizer cooperatives, sugar cooperatives, and dairy cooperatives. The Indian Farmers Fertilizer Cooperative Limited (IFFCO) is one of the biggest manufacturers of fertilizers in the world. The National Agricultural Cooperative Marketing Federation of India (NAFED) is the focal organization of marketing cooperatives for agricultural produce in the country, founded under the Ministry of Agriculture in 1958. It is now one of the largest procurement and marketing agencies for agricultural products in India.

Commodity Boards

Given the vast area and diverse agroclimatic regions, many different crops, commodities, animals and fish species are produced across within India. There are 20 agri-export zones within India. There are five statutory commodity boards under the Department of Commerce. These boards are responsible for production, development and export of tea, coffee, rubber, spices and tobacco. In order to promote other commodities, a number of commodity development boards were established at national and state levels. In most cases, the organizational structure, research, extension and marketing systems are in the process of changing. Detail information on commodity boards is provided in a separate chapter. Thirteen centrally governed commodity boards are listed below.
- Central Silk Board (CSB)
- Coconut Development Board (CDB)
- Coffee Board (CB)
- Coir Board
- Rubber Board (RB)
- Spices Board (SB)
- Tea Board (TB)
- Tobacco Board (TB)
- National Dairy Development Board (NDDB)

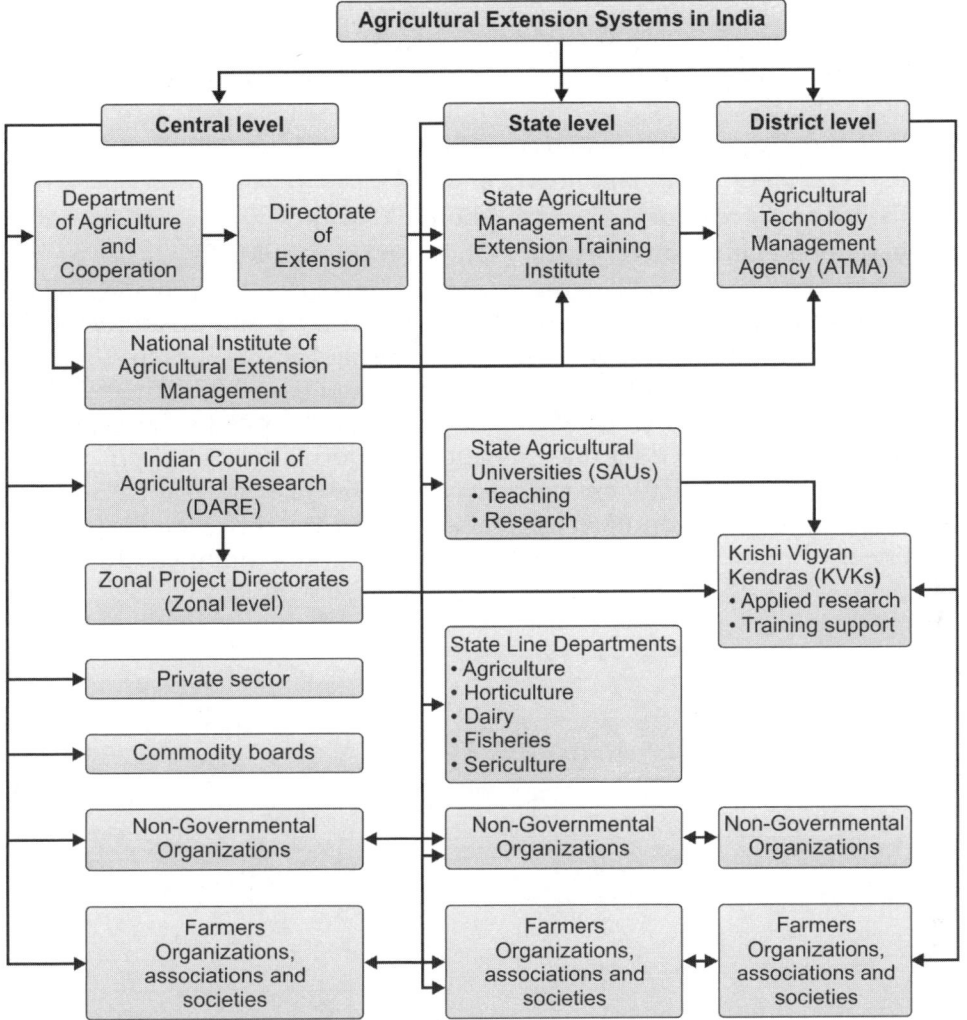

Fig. 3.3: Agricultural extension systems in India. *Source:* Singh KM and Meena MS and Swanson BE and Reddy MN and Bahal R (2014)

- National Horticulture Board (NHB)
- Cashew Export Promotion Council (CEPC)
- National Jute Board (NJB).

National Mission on Agriculture Extension and Technology (NMAET)

The feedback on the performance of ATMA through various evaluation studies and the observations of the working group on extension set up by the Planning Commission to prepare for the Twelfth Five Year Plan paved the way a comprehensive new national scheme to address extension services. The National Mission on Agriculture Extension and Technology (NMAET) was launched by the Department of Agriculture and Farmers' Welfare (DACFW) in 2014-15 and takes a holistic view of extension by embedding components for technical support and training in four major sub-schemes.

It aims to make the system farmer-driven and accountable by restructuring and strengthening existing agriculture extension programmes to enable the delivery of technology and to improve the current agronomic practices of farmers. NMAET consists of 4 Sub-Missions:

a. Sub-Mission on Agricultural Extension (SMAE)

b. Sub-Mission on Seed and Planting Material (SMSP)

c. Sub-Mission on Agricultural Mechanization (SMAM)

d. Sub-Mission on Plant Protection and Plant Quarantine (SMPP)

Sub-Mission on Agricultural Extension (SMAE) aims to focus on creating awareness about the latest technologies to be used by the agriculture and allied sectors. There will be increased training of personnel under agriculture clinics and agriculture business centres. A lot of emphasis has been placed on use of ICT interventions like pico projectors, low cost films, handheld devices, mobile based services, Kisan Call Centres (KCCs), etc. to speed up dissemination of information, good practices, etc.

The guidelines for NMAET were issued well into the financial year 2015-16 and it is too early to say if the new approach has made a significant impact on the ground. It is also pertinent to point out that Government of India has significantly altered the funding pattern in 2015, reducing the Central share from 90% to 10%.

Public and Private Sector Extension System

Extension has been traditionally funded, managed and delivered by the public sector all over the world. Agricultural extension in India has grown over last six decades. It is supported and funded by the national government—through its Ministry of Agriculture (MoA) and other allied ministries. The share of agriculture in Gross Domestic Product (GDP) has declined from over half at the time of independence to less than one-fifth this year. Indian agriculture sector has an impressive long-term record of taking the country out of serious food shortages despite rapid population increase, given its heavy reliance on the work of its pluralistic extension system. The main responsibility for extension activities rests with state governments, since agriculture is the state subject. The central government also implements several technology transfer plans through state governments. Also, Indian agriculture is becoming increasingly more pluralistic in nature, where a large number of private sector firms and civil society extension service providers (e.g. NGOs) co-exist with this public extension system.

PUBLIC SECTOR EXTENSION SYSTEM

In Indian extension system, information flow within public sector moves linearly, with content focusing on transfer of technology for enhancing agricultural production. A wider definition of agricultural extension, beyond improving crop productivity, has not been embraced. Information flow is supply-driven and not need based or area specific (Raabe, 2008), therefore farmers see the quality of information provided by public extension staff as a major shortcoming (NSSO, 2005). There are also insufficient funds for operational costs, training, and capacity development, which limits the activities and continual development of the extension staff (Swanson, 2006). However, it was experienced that there are about 90,000 on the job, which is an adequate number of extension workers for the number of farmers (about 130 million). Various line departments at the state and district levels have been criticized for working in isolation, with weak linkages and rare partnerships. The research–extension link has been criticized for not absorbing or using feedback from farmers and extension staff. Extension personnel and farmers are passive actors, and scientists have limited exposure to field realities (Reddy et al. 2006). Numerous components of public-sector extension system suffer from duplication of programs, without convergence. While

ATMA is pushed as the platform through which the multiple agencies can converge, the implementation difficulties are proving great for effective integration, with shortages of both personnel and funds (Working Group on Agricultural Extension, 2007).

1. *Krishi Vigyan Kendras:* Krishi Vigyan Kendras (KVK) are the field research units of the national agricultural research system (the Indian Council for Agricultural Research-ICAR) and are meant to test new seed varieties, agronomic practices, machinery, etc. in field conditions across different agroclimatic zones before these are cleared for adoption by farmers. The KVK initiative was launched in 1974 and has grown into 611 centres by the end of 2011, ensuring at least one KVK in each district of the country. Besides research, these institutions also conduct farmer outreach programmes through on farm demonstration plots, training, etc. However, despite their impressive network, KVKs are seen as underperformers in terms of reaching out to large numbers of farmers and have faced severe challenges of capacity, performance standards and accountability.

2. *State Agricultural Universities:* The State Agricultural Universities (SAUs) are another important arm for promoting extension activities in the States. While their main mandate is formal degree programmes in major agricultural disciplines, they provide extension and training support through the directorate of extension and education. The information flow is mainly from the universities to the KVKs which are responsible for training farmers. The information flow is largely linear, with little scope for feedback from farmers. Another criticism is that the information flow largely reflects centralised agendas rather than catering to local needs, with the major focus being on transfer of technology. A holistic approach at support to the entire production chain, including post-harvest management is missing.

Information and Communication Technology (ICT) led Extension: An important reform undertaken in recent years by the Ministry of Agriculture at the national level has been the increasing use of modern technologies and communication strategies to educate farmers. ICT has significant potential to reach large numbers of farmers in a cost-effective manner. It can also facilitate two way information flows between farmers and the extension agencies. Here we focus on some of the schemes launched in the past few years.

Farmers Portal: Farmers Portal is a platform where farmers can access information on crop insurance, storage, crop advisories, extension activities, seeds, pesticides, farm machinery, fertilizers, market prices etc. Farmers can download a handbook which provides details of schemes and guidelines of various schemes and programmes.

mKisan: mkisan is an SMS portal that enables authorities at the central and state level to give information to farmers in the local language. There are several free, mobile based applications (or apps as they are commonly referred to), such as KisanSuvidha, PusaKrishi, Agricultural Market, Bhuvan Hailstorm, etc. providing various types of information to farmer through mobile phones.

Kisan Call Centre: These toll-free, phone based agricultural advisory services in local languages are operational in most States with financial assistance provided by Government of India. A single number is offered to farmers for seeking information and advice on a range of agriculture related issues. Subject matter specialists are available at these centres to respond to calls, in case the queries require specialist consultation, a callback facility is also operational. In several States, the KCC has achieved fairly impressive levels of penetration.

Kisan TV Channel: A dedicated 24 hour television channel on agriculture was launched by the national broadcaster, Doordarshan in 2015. Delivering content in several major regional languages, besides Hindi, the Kisan Channel provides real time information on inputs, farming techniques, water conservation, etc. to the farmers. The dedicated channel replaced a daily hour (Krishi Darshan) set aside by the national broadcaster on its terrestrial network of regional centres which produced (in-house developed) programme content on agriculture. Despite criticism by experts for the lack of innovation and attractive production values, Krishi Darshan was listed among the major sources of agricultural advice by farmers (NSSO 70th Round). The majority of those who watched this programme also found the content useful. Government of India seems to have built on this feedback by dedicating a separate channel to agriculture. While initial teething troubles are to be expected, the success of the channel in reaching out to farmers in diverse agroclimatic zones to address their technical needs for a range of crops could be a major factor in influencing growth in the sector.

Agriculture Clinic and Agriculture Business Centres (ACABC): The ACABC scheme was launched in 2002 and was targeted at young rural agriculture graduates who wanted to turn entrepreneurs seeking to provide fee-based agriculture services to farmers. The scheme involves mandatory training and subsidy to set up a rural service centre, often supported by a bank loan. ACABCs were to provide a range of services, including sale of inputs, agriculture advice, marketing support, etc. A mandatory two month training at the National Institute of Agricultural Extension Management (MANAGE), at Hyderabad was designed to install the basis of business management among aspiring agriculture entrepreneurs. As of November 2013, a total of 34,883 graduates were trained under the scheme, 13603 of whom went onto set up agriculture clinics (MANAGE,http://www.manage.gov.in/). However, problems of raising capital and access to institutional finance have proved to be the bane of this programme. The experience so far does not suggest that the intervention is a game changer in the sector and could address the gaps in AES which opened up after the collapse of T & V in many States. During the Twelfth Plan, the scheme was further liberalized in terms of outlays but the impact of these measures can only be assessed after further evaluation.

Major Constraints of these Schemes

- Lack of awareness about the scheme
- Non-cooperation from the banks in promoting agriventures
- Lack of seriousness and attention bestowed on the program by the training institutes
- Poor handholding support by the training institutes
- Absence of dedicated nodal officers at the training institute level for coordinating the scheme
- Inadequate funds for training and handholding activities
- Lack of support from state governments in implementation of the scheme

- Absence of participation of Agri-business companies in implementation of the scheme
- Unattractive credit package for the agri-preneurs for starting agriventures
- Complicated procedures for obtaining license for sale of inputs.

Strategies

- Create awareness about the scheme through electronic media, print media, ATMAs, Banks and Agricultural Universities. Adequate fund may be provided for the scheme for advertisement and publicity.
- Continue the scheme with credit-linked subsidy to complement the public extension system.
- Fund allocation for training and handholding may be increased. Active involvement of ATMA, SAUs, Banks, Agri-business companies and state departments in handholding activities may be strengthened. Operational guidelines are enclosed.
- The upper ceiling limit of credit may be removed. Full interest subsidy should continue to be provided for the first two years.
- The certificate obtained under the scheme by agri-preneurs may defacto be treated as license to sell the inputs.
- Involve Agri-preneurs in extension delivery of the Government extension programme.

Strengthening Agri-Clinics and Agri-Business Centres Scheme

The scheme aims at supplementing the efforts of public extension, to provide specialized extension services to the farmers and to generate self-employment opportunities to unemployed agriculture graduates. The present scheme has provision for training and handholding activities. The impact study conducted by MANAGE (2006) revealed that an agripreneur with 32 months of business experience earns average monthly income of ₹7950/-, covering 38 villages and 3013 farmers through services. Impact of services provided by agriprenuers resulted in increase in yield by 17.4% and income by 28.8%. The response from banks in terms of advancing loans is found to be encouraging.

The scheme would bridge the manpower inadequacy at grassroot level besides making available quality extension services, diagnostic facilities and infrastructure support. The scheme would help the state agriculture and line departments in providing better services to the farmers. Considering long-term benefits of the scheme, more active support of the state governments will help establishment of Agri-Clinics in rural areas.

Banks play significant role in establishment of agriventures. Their involvement during training especially for project preparation would strengthen the acceptability of the projects for credit purpose. Wide network of banks may be used in credit disbursement in definite timeframe.

ATMA's and SAUs can create awareness about the scheme. ATMA may prefer Agri-preneurs in channelizing 10% funds available under Extension Reforms to carry out extension activities at district and below level. This would not only strengthen agriventures but also professionalize the extension services.

Considering the long-term benefits on agricultural extension, it is proposed to continue the scheme with strengthened training, landholding besides credit linked back ended capital subsidy and coupled with interest subsidy. Recently cleared subsidy may be continued to support establishment of agriventures and complement extension.

Agriculture Fairs and Exhibitions: These events have become a common feature in most States and are often effective in demonstrating new technologies and products. Some States, like Gujarat, organize an elaborate fortnight long *Krishi Mahotsav* (which literally means "festival of agriculture"), where mobile vans fitted with video screens and field staff move along a pre-announced route, showing films, holding discussions, selling inputs, etc. These fairs also provide an opportunity for exchange of ideas as well as knowledge and experience among farmers.

Community Radio Stations: Community radio stations are narrow broadcast channels which seek to generate locally relevant content and advice within a small area (typically about a few hundred villages). They are an effective means of dissemination of local knowledge and good practices, help to showcase success stories and mix entertainment, news and other non-technical content along with their core mandate of agriculture extension. While there was much expectation from this medium about a decade ago, the promise seems did not materialize into replicable, scalable models. Perhaps the fact that the bulk of community radio licenses were issued to universities and academic institutions made the content too heavily dependent on experts and academics. The participatory, local nature of community radio could never evolve for this reason. In the interim, more economical means of reaching content to individual farmers (through phones, internet, etc.) overtook radio broadcast technology and the initiative seems to have joined the list of sub-optimal performers in the AES pantheon.

PRIVATE SECTOR EXTENSION SYSTEM

To diffuse agricultural information directly to farmers, private-sector examples are developing context-specific models and using ICT tools. In India, private sector is playing an important task in extension services. The public sector recognizes this, with the policy framework for agricultural extension referring to the need for public extension services not to crowd out private services. Additionally, policy framework for agricultural extension notes that "public extension by itself cannot meet specific needs of various regions and different classes of farmers" (India, DAC, 2000). In the pluralistic extension systems, private sector can provide services related to proprietary goods, while the public sector can provide extension services related to public goods, which tend not to be addressed by private-sector firms. Furthermore, private sector serves a corporate interest, working with individual farmers, so social capital is not built. Moreover, private extension can only work well if farmers are willing and able to pay indirectly through the sale of inputs. It was suggested that private sector could serve the needs of medium-size and commercial farmers, while the public sector could work in remote areas, which are currently not serviced well. This sort of system would require Public-Private Partnership (PPP) that currently does not exist in India. It would mean changes in the way the public sector views and interacts with the private sector. Relying on the public sector may also be difficult for remote and resource-poor farmers, considering the existing problems and poor reach of the public sector in those areas.

AES in private sector are mostly delivered by input dealers, such as those marketing seeds, fertilisers, pesticide and farm machinery. There are about 2.80 lakh input dealers across the country, compared to approximately 1.42 lakh sanctioned posts of extension workers (of which on an average 30% remain unmanned). This gives an idea of the reach and importance of the input dealer as a source of technical advice to farmers. A major complaint against input dealers is that they indulge in "product advisory" instead of "technical advice" which is brand agnostic. Even the Government of India has recognized the leverage of this category of extension support to farmers and offers a course at MANAGE specifically targeting input dealers who wish to brush upon the latest technical knowledge in various sub-sectors of agriculture.

Some private sector agribusiness and input manufacturing companies also undertake direct extension activities. Hyderabad based Nuziveedu Seeds has done a lot of extension related work through its programme, 'Subeej KrishiVignan'. These extension activities are in support of their product brand and seek to help the farmer realize higher production (and thus returns) through necessary pre-sowing preparation, optimum seed rate, correct agronomic practices, application of nutrients and harvesting techniques.

In the case of fertiliser companies, especially large cooperatives like IFFCO (Indian Farmers Fertiliser Cooperative Limited) and KRIBHCO (KrishakBharati Cooperative), extension activities include a wider range of interventions, such as conducting farmer meetings, organizing crop seminars, arranging for soil testing facilities, adopting villages, etc.

Tata Chemicals initiated Tata Kisan Kendras with the objective of empowering and enabling farmers towards improved agronomic practices and higher returns. DCM Shriram, which also produces seeds and fertilizers, established Hariyali Kisan Bazaar (HKB), a chain of agriculture input retail stores which also offered marketing support for select produce. Farmers could also access technical information, information on agri-inputs and banking and farm credit facilities through the HKBs. AGROCEL an agrochemical company, provided inputs and necessary technical guidance to farmers through its "Agrocel Service Centres" in many states. The commercial model adopted by both Tata and DCM Shriram proved unsustainable, leading to closure of the majority of centres initially launched. A similar fate awaited Mahindra ShubhLabh, which was closely modeled on the Tata centres.

ITC, another agribusiness major, launched its e-Chaupal initiative in extension over a decade ago. A VSAT-enabled internet connection at the village level allowed farmers to check prices in the local mandis before they moved their produce for sale. This helped to reduce information asymmetry to a great extent and forced the mandis to adopt fairer price discovery processes. ITC also purchased small quantities of select commodities at these centres for its own trading and processing needs. The e-choupal also provided access to information about weather and innovative farming practices to the farmers. Other initiatives taken by ITC include the "Choupal Saagars" and "Choupal Pradarshan Khet" (CPK). Choupal Saagars mainly comprise of collection and storage facilities which create a hypermarket in rural areas that serves multiple services under one roof. Choupal Pradarshan Khet is a demonstration plot which helps farmers to learn best agronomic practices to enhance their farm productivity. Companies like Pepsico and Heritage Foods, which undertake contract farming of potato and vegetables respectively, also work closely with farmers to provide inputs, technical advice and marketing services (Sulaiman, 2012). None of these models, however, operate at a scale of over a few thousand farmers at the limit, thereby restricting the scope of impact that they make on the wider farming ecosystem.

The growing importance of private sector in both research and extension in India gives rise to an important aspect that has special relevance to the incentives for agriculture research, is that of intellectual property rights (IPRs). A lack of well-defined IPRs weakens incentives for privately funded research.

CIVIL SOCIETY (NGO) EXTENSION SYSTEMS

Within information value chains, the capacity of farmers to articulate their needs will influence their ability to obtain information they need. Considering a large number of marginal and small land holdings in India, both the Farmers Interest Groups (FIGs) and Self Help Groups (SHGs) can play important roles in articulating the needs of men and women farmers to knowledge intermediaries. These FIGs/SHGs can operate side by side with either NGOs or the public sector. However, challenges exist in both sectors (i) Public capacity to build FIGs and SHGs is limited, while NGOs, which are not numerous, rely on donor funds and would need public support to develop the technical skills to facilitate groups, (ii) Building social capital is critical in overall agricultural development strategies for reducing rural poverty (Swanson, 2006). (iii) In a large country like India, through public extension system, meeting of scientists with farmers and visit of farmers to research institutes is a time consuming and difficult task. Both FIGs/SHGs are already emerging as an effective mechanism for both the transfer of technologies and the empowerment of the rural poor (Meena et al. 2003; Meena et al., 2008). Adoption of this approach can reduce the extension cost and workload of extension functionaries. (iv) For that, ICTs could be useful tools to increase connectivity between the various FIGs/SHGs and different extension approaches. Covering the whole country where diversities and complexities are prevalent in agriculture as well as mentally make-up for converting into social capital (especially of the downtrodden, like landless laborers, smallholders, rural women, etc.) is a herculean task. (v) Capacity building of SHGs/FIGs and promoting development of leadership and management skills are utmost needed so that farmers can demand information they need. It is therefore an important component of agricultural extension approaches (Bharati et al., 2014).

In India about 15,000–20,000 NGOs are actively involved in development of rural areas. Their grassroots orientation and proclivity to work in rain-fed and tribal regions has naturally oriented them towards land based livelihoods, hence the essential component of extension in their intervention.

Some NGOs, such as Professional Assistance for Development Action (PRADAN), Bharatiya Agro-Industries Federation (BAIF) and Action for Food Production (AFPRO) are actively involved in promoting extension activities in more than one state. PRADAN has mainly focused on promoting livelihood of the poor in different sectors ranging from agriculture and natural resource management to microenterprise in rural areas across 8 states in India. BAIF is also working on the development of livelihoods by engaging in livestock development, environment conservation, water resource management across 16 states. Syngenta foundation, India (SFI) has been instrumental in helping marginalized farmers adopt high quality production technology for better productivity and improved incomes through unique models of agriculture extension. During 2005-09, three extension-driven projects were launched in targeted disadvantaged regions in the country which included high performing seeds, improved agronomic practices and new pest control technologies. From 2009 to 2013, extension was synonymous with 'market-led extension' especially in the case of vegetables. The idea was to 'produce together and sell together' with fewer intermediaries. In 2014, SFI introduced its flagship initiative of the Agri-Entrepreneur (AE) Model which brought together unemployed youth with an aptitude for entrepreneurial activity to take the role of agri-entrepreneurs. The critical role of an AE is to bring together credit and market linkage, access to high quality input and crop advisory for a 'cluster' of farmers. The AEs must also come from the villages in the cluster they support. Financial support comes from IDBI bank, moreover the AEs are closely monitored and supervised by SFI and partner NGOs to make sure they do not act in self-interest alone. The model is successfully running across six states including three of our study states: Bihar, Madhya Pradesh and Odisha with a network of 40,000 farmers as beneficiaries and a total of 309 AEs as of December, 2017. KRIBHCO, the fertilizer cooperative, launched a not-for-profit entity, called "Gramin Vikas Trust" (GVT), which promotes holistic rural development activities. This organisation is working across 8 states and specializes in the field of agriculture, watershed development, natural resource management, livelihood improvement, women empowerment, institutional development, etc. (Sulaiman, 2012).

Technology Transfer vis-à-vis Extension

TECHNOLOGY

Technology is the making, usage and knowledge of tools, machines, techniques, crafts, systems or methods of organization in order to solve a problem or perform a specific function. It can also refer to the collection of such tools, machinery and procedures. The word "technology" comes from Greek word *techne* meaning "art, skill, craft", and logia, meaning "area of study". The term either can be applied generally or to specific areas: examples include construction technology, medical technology and information technology. Technologies significantly affect human as well as other animal species' ability to control and adapt to their natural environments. The human species' use of technology began with the conversion of natural resources into simple tools.

Dictionaries and scholars have offered a variety of definitions. The Merriam-Webster Dictionary offers a definition of the term, "the practical application of knowledge especially in a particular area" and "a capability given by the practical application of knowledge".

According to Oxford Dictionary, technology is: (1) the application of scientific knowledge for practical purposes, especially in industry, (2) machines and devices developed from scientific knowledge and (3) branch or knowledge dealing with engineering or applied science. Your dictionary defined technology as 'the science or knowledge put into practical use to solve problems or invent useful tools'.

Ursula Franklin (1992) in her "Real World of Technology" lecture gave another definition of the concept; it is "practice, the way we do things around here". Scientific knowledge when put to routine use for benefit of mankind, is called technology (Nigam and Gowda, 1996).

Technology can be most broadly defines as the entities, both material and immaterial, created by the application of mental and physical effort in order to achieve some value. In this usage, technology refers to tools and machines that can be used to solve real-world problems. It is a far reaching term that may include simple tools, such as crowbar or wooden spoon or more complex machines, such as a space station or particle accelerators. Tools and machines need not be material; virtual technologies, such as computer software and business methods, fall under this definition of technology.

The word "technology" can also be used to refer to a collection of techniques. In this context, it is the current state of humanity's knowledge of how to combine resources to produce desired products, to solve problems, fulfill needs, or satisfy wants; it includes technical methods, skills, processes, techniques, tools and raw materials.

Initially, it was applied to arts only, and gradually these "arts" themselves came to be the object of designation. By the early 20th century, the term embraced a growing range of means, processes and ideas in addition to tools and machines. By mid-century, *technology* was defined by such phrases as "the means or activity by which man seeks to change or manipulate his environment". Even such broad definitions have been criticized by observers who point out the increasing difficulty of distinguishing between scientific enquiry and technological activity. "Technology" rose to prominence in the 20th century in connection with the second industrial revolution. The meaning of technology changed in the early 20th century when American social scientists, beginning with Thorstein Veblen, translated ideas from the German concept of "technik" into "technology". In German and other European languages, a distinction exists between Technik and Technologie that is absent in English, as both terms are usually translated as "technology". By the 1930s, "technology" referred not to the study of the industrial arts, but to the industrial arts themselves. In 1937, the American Sociologist Read Bain wrote that "technology includes all tools, machines, utensils, weapons, instruments, housing, clothing, communicating and transporting devices and the skills by which we produce and use them"(Wikipedia).

From a historical perspective, philosophers of technology agree that two phases of technology can be seen: the craft phase and the modern scientized phase. However, to a philosopher of technology, modern technology, although scientized, is a unique structure of thinking, not merely applied science. Nor, is technology, like science, fully described by the laws of nature. There are many lessons and best practices that can be gleaned from existing studies if technology is looked at in broader terms. Gershon and Umali defined technology as "… a factor that changes the production function and regarding which there exists some uncertainty, whether perceived or objective (or both). The uncertainty diminishes over time through the acquisition of experience and information, and the production function itself may change as adopters become more efficient in the application of the technology" (Parvan, 2011).

It is the application of science and technical advance to the production of materials to serve human needs. The concept of technology can be defined simply to mean the application of science in the development, utilization or application of materials or things or methods of undertaking certain activity or work (Hashim, 1978). According to Yotopoulos and Nugent (1976) technology is a body of knowledge that can be applied in productive process. According to Rogers (1962), a technology usually has two components: (1) a hardware aspect consisting of the total tool that embodies the technology as materials or physical object and (2) a software aspect, consisting of the information base for tool.

Venkatasubramanian et al. (2009) defined technology as any systematic knowledge and action applicable to any recurrent activity. Technology involves application of science and knowledge to practical use, which enable man to live more comfortably.

AGRICULTURAL TECHNOLOGY

Agricultural technology focuses on technological processes used in agriculture, to create an understanding of how processes, equipment and structures are used with people, soil, plants, animals and their products, to sustain and maintain quality of life and to promote economic, aesthetic and sound cultural values (http://www.fao.org/sd/teca/def_en.asp,dfid-agriculture-consultation.nri.org).

Agricultural technology refers to new inputs, methods, new process or new innovation to increase the production and productivity in agriculture. Agricultural inputs like land, labour and capital are used to produce agricultural products like vegetable, crop. The transformation process is guided by technology. Thus, the level of agricultural production or productivity is guided by the technology we are using in farming. (www.agricommercialization.blogspot.com).

Samanta (1985) defined agricultural technology as a body of systematically organized knowledge and materials applicable to local production problems to help to boost the present level of productivity and or extend the existing range of production.

Kumar and Hansra (2001) identified the following distinguish features of new agricultural technologies:

- They are input intensive and expensive.
- They are market oriented rather than consumption based.
- They require complex technical knowledge and skills.
- They require network of institutional arrangement.
- They are more vulnerable to insect, pest and diseases.

According to Conway and Barbier (1990), the potentiality of agricultural technology may be assessed through production system properties like productivity, stability, sustainability and equitability. Agricultural technology has been a primary factor contributing to increases in farm productivity in developing countries over the past half-century. Although there is still widespread food insecurity, the situation without current technology development would have been unimaginable.

As conceived in DFID working paper 4, agricultural technology assume to include: the products of plant and animal breeding (including biotechnology); the introduction of new crops; improved management practices relating to crops, livestock and fisheries; mechanization; infrastructure development; external inputs (including chemicals, bio-control products and veterinary products) and local inputs (soil amendments, mulches, etc).

Types of Agricultural Technology

Grabowski, Sivan and Tracy (1986) pointed out two types of agricultural technology: (i) mechanical and (ii) biochemical. Mechanical technology involves the application of machinery to the production process, i.e. tractor, thrasher, irrigation pumps, etc. Some part of it results in increased yield. However, for the most part it is thought that, that type has little impact on yields. Biochemical technology is generally yield increasing and is really package of inputs: seeds, fertilizers and irrigation water. Two types are interdependent of each other in terms of application.

In agriculture, the term technology often confuses practitioners. This is because agricultural technology is a complex blend of materials, processes and knowledge. Swanson (1997) has classified agricultural technologies into two major categories:

1. Material technology, where knowledge is embodied into a technological product.

2. Knowledge based technology, such as the technical knowledge, management skills and other processes that farmers need for better farm management and livelihood support.

Scale Neutral Technology

According to Rudra (1982), technology can be called scale neutral if the responses to divisible inputs such as water, fertilizer, seed, etc. are not found to depend on the size of plot.

Grabowski (1967) stated the technology is scale neutral if it can be effectively and efficiently used in small as well large farm; otherwise, there exists scale bias.

Appropriate Technology

The concept of appropriate technology (AT) stemmed from the work of British economist Dr. Fritz Schumacher in the 1970s. Appropriate technology is a grass roots approach to technology that builds a strong sense of community and encompasses benefits that span across social, environmental, cultural, economic and spiritual facets. Appropriate technology is not a one size fits all approach, but rather adapts to best fit with the community in which it is developed. Appropriate technology best fits with the community it serves because it is created by the people to meet a need. Therefore, the communities are placed at the centre of decision-making and create technologies that will best serve their communities in the long-term (http:// web.uvic.ca/-essa/ wp-content/uploads/2010/03/Reclaiming-Sustainability-Conference-appropriate technology.pdf).

According to Merriam Webster Dictionary, technology is suitable to the social and economic conditions of the geographic area in which it is to be applied, is environmentally sound, and promotes self-sufficiency on the part of those using it. As stated in Macmillan Dictionary technology that is suitable for the place in which it will be used, usually involving skills and materials that are easily available in the local area.

A science or technology considered reasonable and suitable for a particular purpose that conforms to existing cultural, economic, environmental and social conditions. Appropriate technology is economically viable, regionally applicable, and sustainable (Dictionary.com).

According to Appropedia, "Appropriate Technology (AT) is technology that is designed to be "appropriate" to the context of its use. The most appropriate technologies are:

- Sustainable-requiring fewer natural resources and producing less pollution than techniques from mainstream technology, which are often wasteful and environmentally polluting.
- Small where possible (as in small is beautiful). This places more power at the grass roots, in the hands of the users. However, there are also times when the most appropriate technologies are large-scale.
- Appropriate to the context, including the environmental, ethical, cultural, social, political and economically context. The appropriate technology for one context may not be appropriate for another.

All the research findings are not technology, the research findings which are having some practical utility, tested, and found economically feasible, socially acceptable, culturally adaptable, scale neutral, gender neutral and eco-friendly at the farmers conditions are considered as appropriate farm technology. According to Singe (1978), a piece of technology may be viewed as appropriate for a society if its design was concerned with real needs of that society in mind, its use fulfill these needs, its continuance and development are based on the society's economic and technical ability to support, service, maintain even improve upon it.

Management

As defined Oxford Dictionaries management is the process of dealing with or controlling things or people.

As stated in Business Dictionary, management could be the organization and coordination of the activities of a business in order to achieve defined objectives. Management is often included as a factor production along with machines, materials and money.

It is very difficult to give a precise definition of the term 'management'. Different scholars from different disciplines view and interpret management from their own angles. The economists consider management as a resource like land, labour, capital and organization. The bureaucrats look upon it as a system of authority to achieve business goals. The sociologists consider managers as a part of the class elite in the society.

As cited in http://www.whatishumanresource.com/management-definitions-by-great-management-scholars; various definitions of management given by different management authors are:

- According to George R. Terry, *"Management is a distinct process consisting of planning, organizing, actuating and controlling; utilizing in each both science and art, and followed in order to accomplish pre-determined objectives"*.
- Management is a multipurpose organ, that manage a business and manages, workers and work (Peter Ferdinand Drucker).
- According to Peterson and Plowman, "Management may be defined as the process by means of which the purpose and objectives of a particular human group are determined, clarified and effectuated".
- One popular definition is by Mary Parker Follett, Management, she says, management is the "art of getting things done through people".
- According to Harold Koontz, "Management is the art of getting things done through others and with formally organized groups".
- According to FW Taylor, "Management is the art of knowing what you want to do and then seeing that they do it in the best and the cheapest may".

Technology Management

The US National Research Council in Washington DC, defined management of technology (MOT) as linking "engineering, science and management disciplines to plan, develop and implement technological capabilities to shape and accomplish the strategic and operational objectives of an organization" (National Research Council, 1987). While technology management techniques are themselves important to firm competitiveness, they are most effective when they complement the overall strategic posture adopted by the firm.

Technology management can be defined as the integrated planning, design, optimization, operation and control of technological products, processes and services (Venkatasubramanian et al., 2009).

Technology Development

Technology development (also called technology innovation) is a process consisting of all the decisions and activities which a scientist does from a recognition of a need/problem with planning, testing, conducting research, verification, testing and dissemination for adoption. During the same time, some problems on the technology might get back to the scientist for solution thus resulting in refinement of the same. Thus, technology development is a continuous process (Venkatasubramanian et al., 2009).

According to Nigam and Gowda (1996), technology development is an on-going response of scientific knowledge to changing requirements of society. If focuses on a target group keeping mind the resource base, sociocultural factors and government policies to exploit the available opportunities and match scientific knowledge with requirements.

A good development process should be flexible and offer options to the target groups for successful adaptation and adoption of new technology. Generation of new scientific knowledge is essential to upgrade the existing technology, so a strong and well-focused research programme is a prerequisite for any technology development.

According to Venkatasubramanian et al. (2009), technology development basically constitutes seven processes. They are:
1. Technology generation
2. Technology testing
3. Technology adaptation
4. Technology integration
5. Technology demonstration
6. Technology dissemination
7. Technology adoption

Technology generation, the starting point of technology development process is mainly a function of research system. Testing, adaptation and integration processes constitute technology assessment and refinement which KVK system executes through OFTs. The feedback is passed over to research system. KVK system also involves in technology demonstration through FLDs. Feed-forward from successful OFTs and FLDs is communicated to the extension system for mass popularization in the district. Technology adoption; the final act, occurs among the members of client system, i.e. farmers.

The main objective of agricultural research is to solve the farm and farming related problems of farmers by developing appropriate technologies. Research management primarily involves perception/identification and articulation of the research problem, project prioritization, selection and resource allocation, planning of research activities, monitoring and review of the research results, validation, refinement, treatment and

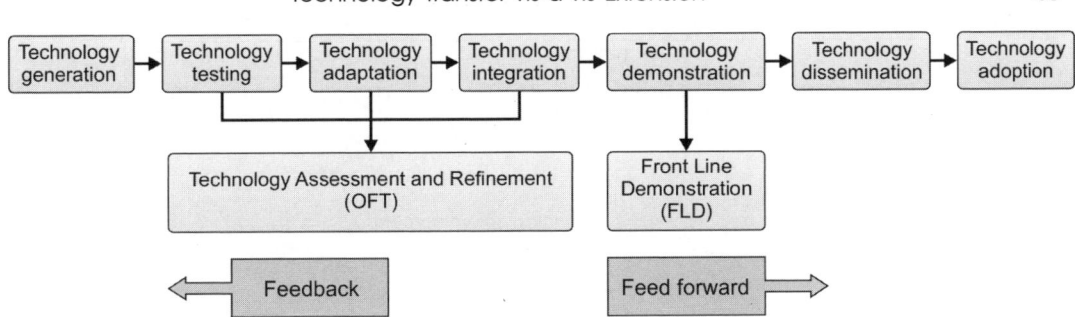

Fig. 5.1: Conceptual model of technology development process. *Source:* Sajeev and Venkatasubramanian (2009)

dissemination of research results (transfer of technology) and utilization of research results by the clients. The research management process consists of various activities centered on developing knowledge and technology and transferring the same to the framing communities to achieve the objectives of agricultural development (Samanta and Sontakki, 2006).

A conceptual model of technology development process in agriculture depicting the components actors involved is explained in Fig. 5.1.

Designing Agricultural Research and Extension Systems

The basic consideration in designing any research-extension system is technology flow. Knowledge of the process of technology flow facilitates diagnosis of research-extension linkage problems. The technology flow concept is based on the premise that technology is derived from science and flows from research station to users with or without an intermediary agency such as an extension service. In agriculture, the term 'technology' is used broadly to also include improved crop varieties and livestock, chemical inputs, farm implements and farm practices (Javier, 1989).

Technology Flow Processes

Technology flow involves sequential processes along the science-practice continuum. They are: science, technology generation, technology testing, technology adaptation research, technology integration, dissemination, diffusion and adoption (Roling, 1989). The transfer of technology model (Chambers, 1993) is most common, where breakthroughs developed by researchers are transferred to extension for delivery to users. This is a one-way, linear process and similar to the progressive farmer approach (Roling, 1989). This assumption of a linear, sequential flow of technology has been criticized by many social scientists as it ignores the actual contribution and potential of farmers as generators of technology (Javier, 1989). The model also neglects policy-driven, market-driven, and farmer-driven innovation.

Several other models—such as the technology innovation process (McDermott, 1987); the research-extension process (Bernardo, 1986); the technology generation and delivery process; and the agricultural technology development system have been developed to describe technology flow. These models have been synthesized in the research-extension interface model (Javier, 1989) shown in Fig. 5.2. The components of this model are: basic research, strategic research, technology generation (applied

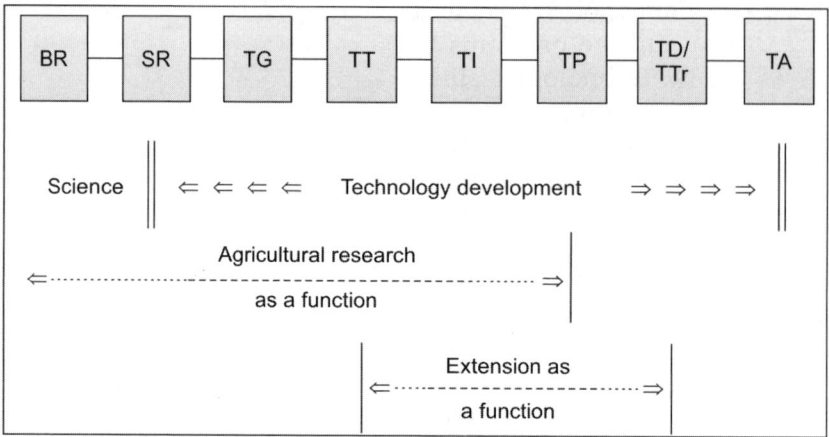

BR = Basic research; SR = Strategic research; TG = Technology generation;
TT = Technology testing; TI = Technology integration; TP = Technology Production;
TD/TTr = Technology dissemination and technology transfer; TA = Technology adoption

Fig. 5.2: The research-extension interface in the technology flow process. *Source:* Javier, 1989

research), technology testing, technology integration, technology production, technology dissemination (technology transfer) and technology adoption. In this model, basic research refers to research conducted in the basic sciences. Strategic research, which mainly includes research on directly applicable basic knowledge, is taken as the boundary between science and technology.

Technology generation, testing, integration, production, dissemination and adoption constitute the technology development process. Technology generation has the same function as applied research, where the knowledge accumulated from basic science research is organized, interpreted, reformulated and translated into technology. Technology testing refers to verifying the results of new technology in the field to obtain early feedback. This is indeed a part of farming systems research (FSR) and on-farm trials. For the purpose of testing, participation of the extension service has been increasingly sought as its widespread networks help in reaching out to farmers, especially in less well endowed regions.

Technology integration involves fine tuning and packaging of technologies into specific recommendations. Technology production involves designing and producing information materials, improved inputs, training programs, etc. Technology dissemination is the process of delivery of technology to farmers.

Agricultural research includes activities from strategic research to technology production, and the extension function includes technology testing to technology dissemination. Thus technology testing, integration and production constitute the essential overlap area for linkage between research and extension.

The Linkage Problem

Agricultural research institutions usually concentrate their effort on strategic research and technology generation. Some efforts towards technology testing are also made. However, technology integration and production activities are generally neglected. In contrast, most extension agencies concentrate their effort on technology production and technology dissemination, with negligible attention given to technology integration and testing. Critical linkage problems therefore arise at the technology integration stage, followed by the technology testing and production stages.

In the absence of clear directions, research or extension personnel concentrate their effort on those activities which, in their judgment, are considered important. The judgment of these personnel are influenced principally by their background, experience and training. Usually, their background and training has not emphasized linkage activities. Often the institutions themselves give a low priority to linkage, especially when research and extension activities are administered by separate institutions. As a result, research institutions and personnel consider strategic research and technology generation to be their 'main activities,' while extension institutions and personnel consider technology production and dissemination to be their 'main activity.' In the process, linkage activities are neglected, or considered subsidiary by both. However, linkage activities cannot be performed in isolation; they require coordination of personnel from both research and extension functions, which demands additional efforts from both functions. Since background and training generally do not emphasize linkage activities, more effort is needed to build expertise in such activities. Thus, the additional effort needed for coordinating and building expertise is also a major constraint when considering these activities to be 'main' for both research and extension personnel. The linkage problem is more severe in cases where research and extension functions are performed by separate public institutions.

In the absence of effective linkage, researchers do not receive enough information about the environment and resource constraints under which farmers are operating. This is an important piece of information for research institutions when setting their priorities and goals. Also, extension agents do not receive the necessary information and cooperation they need from researchers to adapt and then disseminate new technology. The linkage problems thus cause disruptions in technology flow and lead to low adoption rates, increased time lags between development and adoption of new technology, reduced efficiency in the use of resources, unnecessary competition and duplication of efforts, and increased cost of agricultural research and extension activities.

TECHNOLOGY ASSESSMENT AND REFINEMENT (TAR)

Technology assessment and refinement (TAR) refers to a set of procedures whose purpose is to develop recommendation for a particular agroclimatic situation/ location through assessment and refinement of recently released technology through participatory approach. It refers to the process or a set of activities before taking up new scientific information for its dissemination in a new production system. OFTs conducted by KVKs are based on this concept and thus distinguish it from agronomic and research trials. As detailed earlier, the process of TAR has three components. They ate technology testing, technology adaptation and technology integration. TAR should be:

• Site specific
• Holistic
• Farmer participatory
• Technical solution to existing problems

- Interdisciplinary
- Interactive.

This process involves scientific-farmer linkage in terms of:
- Sufficient understanding of the farming situations
- Adequate perception of farmers' circumstances and their needs
- The variability of conditions on the research status as compared to farmers' fields
- Problem orientation instead of disciplinary approach.

Technology assessment in agriculture is the study and evaluation of new technologies under different micro locations. It is based on the conviction that new discoveries by the researchers are relevant for the farming systems at large, and that technological progress can never be free of implications. Also, technology assessment recognizes the fact that scientists at research stations normally are not trained field level workers themselves and accordingly ought to be very careful while passing positive judgments on the field level implications of their own, or their organization's new findings of technologies (Venkatasubramanian et al., 2009).

DISSEMINATION

According to Oxford Dictionary, Dissemination is: The act of spreading something, especially information, widely; circulation: e.g., dissemination of public information.

According to Merriam Webster Dictionary disseminate means to cause (something, such as information) to go to many people. It means: to spread abroad as through sowing seed <disseminate ideas> or to disperse throughout.

It is believed to be originated from Latin *Disseminatus*, past participle of disseminare, from dis-+ seminare to sow, from semin-, semen seed; around 1566 AD.

Meaning and Significance of Dissemination

Dissemination takes on the theory of the traditional view of communication, which involves a sender and receiver. The traditional communication view point is broken down into a sender sending information, and receiver collecting the information processing it and sending information back, like a telephone line.

The information is sent out and received, but no reply is given. The message carrier sends out information, not to one individual, but many in a broadcasting system. An example of this transmission of information is in fields of advertising, public announcements and speeches. Another way to look at dissemination is that of which it derives from the Latin roots, the scattering of seeds. These seeds are metaphors for voice or words: to spread voice, words and opinion to an audience.

Technology Transfer

Technology transfer and diffusion are two aspects of technology dissemination which is the process by which innovations are transferred from donor to receiver. Technology transfer involves communication between a specific recipient and a group of recipients. In technology diffusion the donor is not necessarily aware who the recipients are.

Technology can take the form of an object or a concept or a technique. Technology transfer is the major component of international policy. Adoption and diffusion are the process governing the utilization of innovation. Measures of adoption may indicate both the timing and the extent of new technology utilization by individuals. Adoption behavior may depict by more than one variable, both discrete and continuous (Surling and Zilbermann, 2000).

Defining technology is paramount because it helps to identify phenomena related to technology transfer. Since the 1960s, many scholars have tried to understand the real meaning of technology using different underlying philosophers (Devore, 1987; Frey, 1987; Galbraith, 1967; Mitcham, 1980; Skolimowski, 1966). The definitions or meanings of technology these authors proposed were unique, according to their context, philosophy, economy, or other variables. This implies that it might not be that simple to define technology because technology is situation and value specific. However, the concept of technology should be outlined in order to understand what is being transferred in a technology transfer process.

Two approaches have been used to comprehend technology: is to define technology in a way of capturing the platonic essence in a few sentences by differentiating technology from science, and the other is to provide characterizations of technology. Scholars, such as Skolimowski (1966), Galbraith (1967) and DeVore (1987) might be the representatives of the former approach. Skolimowski (1966) defined technology as a form of human knowledge and a process of creating new realities. He argued that science is concerned with what is, but technology is concerned with what is to be. Later, Galbraith (1967) defined technology as the systematic application of scientific or other organized knowledge to practical tasks. This definition is notable because it emphasizes the systematic and practical aspects of technology. DeVore (1987), a major scholar, made an effort to define technology. He argued that technology should create the human capacity to "do", and it should be used to create new and useful products, devices, machines or systems. He also emphasized the relationship between technology and social purpose. He contended that technology has always been situated directly in the social milieu and conditioned by values, attitudes and economic factors; thus, the goal of technology is the pursuit of knowledge and know-how for specific social ends.

Markert's (1993) definition of technology transfer is the most typical—she defined technology transfer as the development of a technology in one setting that is then transferred for use in another setting. However, this definition does not reflect a deep comprehend of technology transfer, because it is mostly focused on differentiating technology development from utilization. To overcome the disadvantage of this definition, Johnson, Gatz and Hicks (1997) tried to interpret technology transfer through a holistic perspective that included both the movement of technology from the site of origin to the site of use and issues concerning the ultimate acceptance and use of the technology by the end user. They argued that recognizing the end user's needs and the context where the technology will be used is essential for the successful transfer of technology. Technology transfer is not the same process and perception of everybody. Universities, corporations, federal labs and developing countries have different roles and interests in technology transfer. For examples, universities, as a provider of technology, view technology transfer as a means for serving a community through knowledge sharing. On the other hand technology transfer is regarded as a way to obtain competitive advantages through performance improvements in corporations that are the recipients of this technology. Like this, the perception of technology transfer in each site would be different. According to Frey (1987), technology can be an object, a process or knowledge that is created by human intention. In most cases technology tends to be the integration of all three components: object, process and knowledge. Therefore, a provider of technology should try to transfer the integration of all components that make up that technology, not just one component.

The literature on technology diffusion, in general, suggests that the term refers to the spreading, often passively within a specific technological population, of technological knowledge related to a specific innovation of interest to that population. Technology transfer, on the other hand, is a proactive process to disseminate or acquire knowledge, experience and related artifacts (Hameri, 1996). The term technology transfer can be defined as the process of movement of technology from one entry to another (Souder et al., 1990; Ramanathan, 1994).

The transfer may be said to be successful if the receiving entity, the transferee, can effectively utilize the technology transferred and eventually assimilate it (Ramanathan, 1994). The movement may involve physical assets, know-how and technical knowledge (Bozeman, 2000).

Technology transfer in some situations may be confined to relocating and exchanging of personnel (Osman-Gani, 1999) or the movement of a specific set of capabilities (Lundquist, 2003). Mittleman and Pasha (1997) have attempted a broader definition where they stated that technology transfer is the movement of knowledge, skill, organization, values and capital from the point of generation to the site of adaptation and application.

Technology transfer has also been used to refer to movements of technology from laboratory to industry, developed to developing countries or from one application to another domain (Philips, 2002). Even though technology transfer is not a new business phenomenon, the considerable literature on technology transfer that has emerged over the years agrees that defining technology transfer is difficult due to the complexity of the technology transfer process (Robinson, 1988; Spivey et al., 1997). The definition depends on how the user defines technology and in what context (Chen, 1996; Bozeman, 2000).

In a very restrictive sense, where technology is considered as information, technology transfer is sometimes defined as the application of information into use (Gibson and Rogers, 1994). According to Rogers (2003), diffusion is the process by which an innovation is communicated through certain channels over time among the members of a social system and by which alteration occurs in the structure and function of a social system as a kind of social change. Diffusion is an extremely critical process for the practical use of innovation and reinvention. In other words, diffusion plays a pivotal role in helping the adopters fully take advantage of an innovation and to modify that innovation. Thus, the comprehension of the major issues in the diffusion process is essential for making technology transfer successful.

Interest in technology transfer goes back to over six decades. During the colonial era, technology transfer by colonial powers to production entities in their colonies was mainly in the primary sector such as mining, plantation and agriculture (Ramanathan, 1989). Those transfers were aimed at the development of methods and techniques in order to obtain the maximum output in export industries such as mining and plantation agriculture and the development of infrastructure for such industries.

Mansfield (1975) classified technology transfer into vertical and horizontal technology transfer. Vertical transfer refers to transfer of technology from basic research to applied research to development and then to production respectively and horizontal technology transfer refers to the movement and use of technology used in one place, organization or context to another place, organization or context. Souder (1987) refers to the former as internal technology transfer and the latter as external technology transfer. Souder further elaborates upon vertical technology transfer as a managerial process of passing a technology from one phase of its life cycle to another. The importance of technology transfer, from an economic and competitiveness perspective, has also stimulated university-industry technology transfer. This is evident from the emergence of technology transfer offices in most research offices and universities (Siegel et al., 2004).

TECHNOLOGY TRANSFER MODELS

There are many popular models of technology transfer; examples include the appropriability model, the dissemination model, the knowledge utilization model, the contextual collaboration model, the material transfer model, the design transfer model and the capacity transfer model (Rogers, 2003; Ruttan and Hayami, 1973; Sung and Gibson, 2005; Tenkasi and Mohrman, 1995). According to the appropriability model, purposive attempts to transfer technologies are unnecessary, because good technologies sell themselves. Regarding the dissemination model, the perspective is that transfer processes can be successful when experts transfer specialized knowledge to a willing recipient. The knowledge utilization model emphasizes strategies that effectively deliver knowledge to the recipients. A contextual collaboration model is based on the constructive idea that knowledge cannot be simply transmitted, but is should be subjectively constructed by its recipients. The material transfer model focuses on the simple transfer of new materials, such as machinery, seed, tools and the techniques associated with the use of the materials. According to the design transfer model, transfer of designs, such as blueprints and tooling specifications, should accompany the technology itself for effective technology transfer. The capacity transfer model emphasizes the transfer of knowledge, which provides recipients with the capability to design and produce a new technology on their own.

A country's competitive advantages increasingly lie in its capabilities to generate further innovations and to use effectively new technology, which is generally a function of the capacity of its population to absorb new technologies and incorporate them into the production process (Kolfer and Meshkati, 1987). This implies that a successful transfer of technology has a large impact on the advancement of a nation and it significantly depends on the capacity of people to assimilate, adapt, modify and generate new technology. Consequently, educational infrastructure to develop "human capital" is the basic component for a successful technology transfer. After accumulating a high quality of human capital, a recipient of technology should develop an elaborate strengthening the collaboration between the donor and the recipient. Lastly, the recipient should be able to generate new innovation based on the successful transfer of technology. This model can be shaped to increase the willingness of both the recipient and the donor of technology transfer. This plan could facilitate the transfer of technology by strengthening the collaboration between the donor and the recipient.

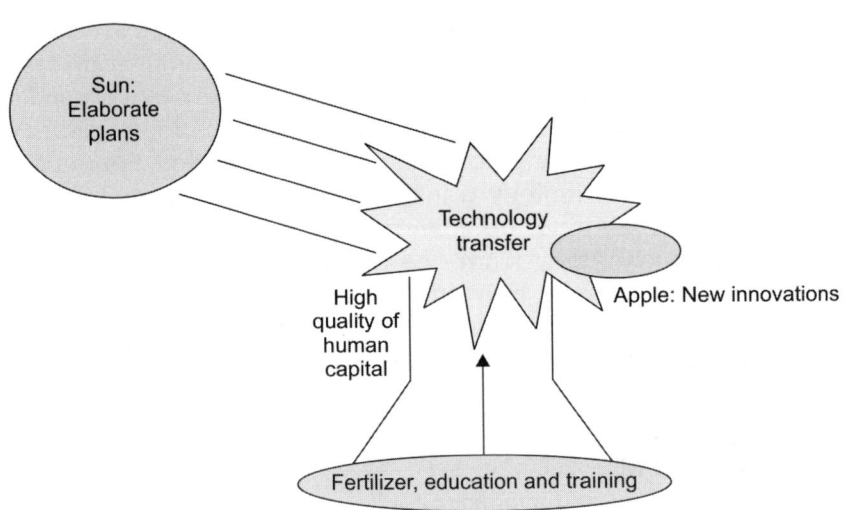

Fig. 5.3: The role shifting model of technology transfer. *Source:* Adopted from (Choi, 2009)

Lastly, the recipient should be able to generate new innovations based on the successful transfer of technology. This model can be shaped as shown in following Fig. 5.3.

The Fig. 5.3 is titled "the role shifting model of technology transfer" because its ultimate goal is to generate new innovations. The model depicts how recipients of technology today can be tomorrow's donor of technology: it shows the conditions that enable fruit to ripen or in other words, new innovations. Thus, a high level of continuing education and training results in the role of fertilizing or helping an apple tree (technology transfer) grow well. In addition, elaborate plans for collaboration between recipients and donors help to achieve successful technology transfer as either sun or rain is helpful for the growth of a tree. Consequently, farmers who are recipients of technology will be able to produce a plenty of fruits (new innovations) based on a high level of continuing education and training (fertilizer) and elaborate plans that play a role of sun and rain.

In practice, extension organizations everywhere pursue the overall goals of technology transfer and human resource development, though the emphasis will differ. Within each organization there is a mix of objectives and within countries there is often a mix of organizational patterns. The model presented in Fig. 5.4, stakeholders and agents in agricultural technology transfer model.

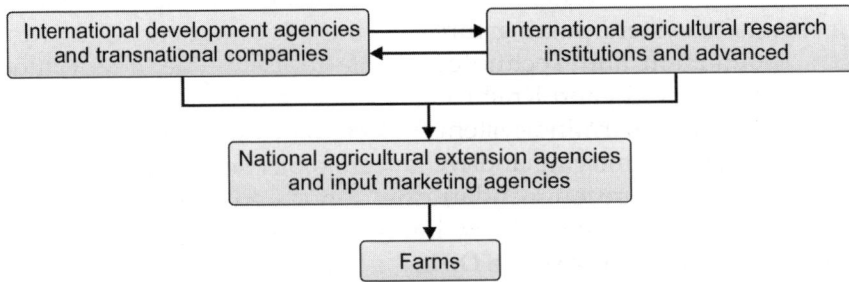

Fig. 5.4: Stakeholders and agents in agricultural technology transfer model. *Source:* Adopted and modified from Thrupp et al. (2000).

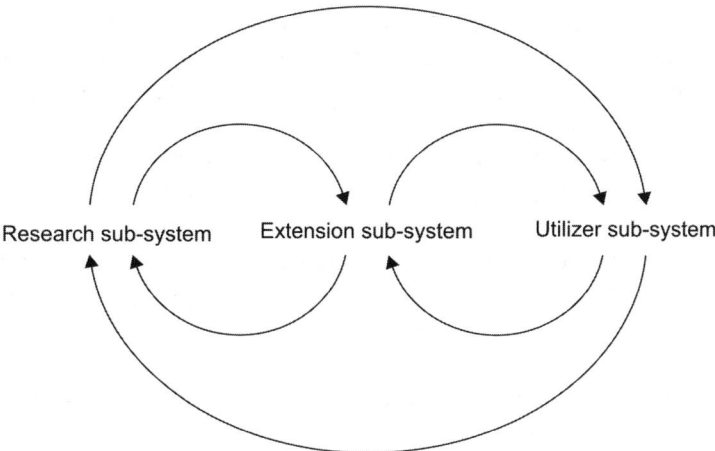

Fig. 5.5: Agriculture knowledge and information system. *Source:* Christoplos and Nitsch (1993) Adopted from Roling (1988).

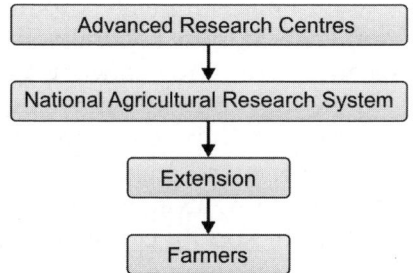

Fig. 5.6: Transfer of technology model. *Source:* Chambers and Jiggins (1986)

This TOT model in Fig. 5.6 is being eclipsed by newer models which acknowledge the overlapping of researchers, outreach workers and farmers (Christophos and Nitsch, 1993 in Fig. 5.5) rather than focusing on the technology itself, the new system recognize that information and knowledge provide a common denominator among farmers, extension workers and researchers.

TECHNOLOGY DISSEMINATION AND EXTENSION STRATEGY

Technology transfer system passing through the stages of Technology Assessment, Technology Refinement and Technology Integration through Institution Village Linkage Programme (IVLP) and Krishi Vigyan Kendras (KVKs). Analysis of cost risk return structure of major farming systems in agro-eco regions is called for. These has to be consensuses on the unifying and well tested recommendations to the farmers in order to avoid conflicting recommendations passing through pipeline.

AGRICULTURAL KNOWLEDGE AND INFORMATION SYSTEM (AKIS)

Agricultural knowledge and information systems (AKIS) comprise the institutions and organizations that generate and disseminate knowledge and information to support agriculture production, marketing, and post-harvest handling of agricultural products and management of natural resources. Most AKIS projects support agricultural research, extension, or education activities, which are increasingly viewed as components of an interrelated system.

An AKIS is a system that links rural people and institutions to promote mutual learning and generate, share and utilize agriculture-related technology, knowledge and information. The system integrates farmers, agricultural educators, researchers and extensionists to harness knowledge and information from various sources for better farming and improved livelihoods. This integration is suggested by the "knowledge triangle".

AKIS systems that are financially, socially and technically sustainable; relevant and effective processes of knowledge and technology generation, sharing and uptake; AKIS systems that are demand-driven through empowerment of farmers such that programmes and activities are responsive to their needs; the interface between and integration among the various education, research, extension and farming activities; and accountability to assure that stakeholders assume their respective responsibilities.

The AKIS concept and practice hold significant promise for the advancement of agricultural and rural development and more generally, national economies. In order to realize the value and importance of the AKIS concept, agricultural institutions need actively to promote linkages, technology transfer, knowledge sharing and the exchange of relevant information and such an impetus to the development of pluralistic innovation systems must be supported by adequate financial commitment. Fundamental to the development of an AKIS is recognition of the role of a plethora of private sector actors (seed and input supply companies, produce buyers, chemical companies, radio and television, etc.) playing different roles within the system.

Rural people, especially farmers, are at the heart of the knowledge triangle. Education, research and extension are services-public or private-designed to respond to their needs for knowledge with which to improve their productivity, incomes and welfare and manage the natural resources on which they depend in a sustainable way. A shared responsiveness to rural people and an orientation towards their goals ensures synergies in the activities of agricultural educators, researchers and extensionists. Farmers and other rural people are partners within the knowledge system, not simply recipients.

An Emerging Approach for Sustainable Development (Saha, 2011) reported that there is conceptual progression from looking at various institutions and practices such as farming system development, extension and research in isolation to considering the linkages between the pairs of these elements as an AKIS. AKIS could involve providing farmers with a basket of opportunities and helping them to choose the right opportunity for their situation.

The AKIS/RD Vision and Principles

- To set forth a shared vision for an integrated approach to agricultural education, research and extension that would respond to the technology, knowledge and information needs of millions of rural people, helping them to reach informed decisions on the better management of their farms, households and communities.

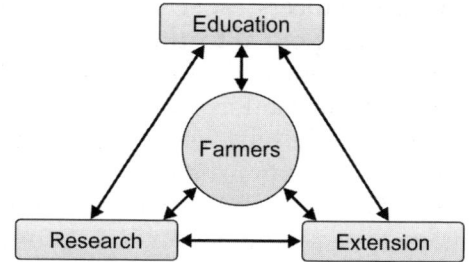

Fig. 5.7: Model of AKIS. *Source:* Sharma, 2012

- To facilitate dialogue with decision-makers, in both government and development organizations, ensuring that proposals for investment in AKIS are well founded and receive due consideration.
- To provide the staff of FAO and the World Bank, and their counterparts in client countries, with a common set of principles to guide their work in agricultural education, research and extension.
- To ensure synergies from complementary investments in education, research and extension, resulting in more effective and efficient systems.

An idealized AKIS model: This model (Fig. 5.7) suggests the numerous elements in AKIS. In fact, it could be made messier in that the policy, physical and human resources, communications and institutional elements should be connected to each of the four main sets of institutions-research, extension, education and support systems—which themselves should include both public and private sector entities.

Challenges of AKIS

- Most poor people depend on agriculture
- Food needs call for steady growth in agricultural production
- Improving rural incomes and raising agricultural production will often require agricultural intensification
- Agricultural intensification must be balanced with environmental sustainability
- Rural people also look to knowledge and information systems for guidance on how to bring about general improvements in their livelihoods
- For people living in this environment, knowledge is key
- Many farmers fail to benefit from technological and other advances
- The AKIS institutions have not been responsive enough in addressing the problems and opportunities facing farmers.

New Opportunities Exist for raising AKIS Effectiveness

Advances in the agricultural sciences are crucial but other advances are also needed.
1. Communication and information technologies are advancing rapidly
2. New concepts are emerging for participation in learning and problem-solving.

Government also needs to create the conditions necessary for developing AKIS. Investment in market development and support to input providers, especially credit and supply institutions, are needed to stimulate the agricultural community, and attention to the rural physical infrastructure is needed to make the environment attractive and safe. Agricultural producers, especially women and poor farmers, require education and training to bring them into the modern world of labour-saving technologies and more productive practices. Joint planning between producers and institutional operators can provide the platform for advancing a demand-driven system of technological innovation for agricultural development. System leaders and managers need a better understanding of the dynamic nature of both national and international technology systems, and should be able to identify those areas where the public system has a comparative advantage over private sector R & D firms. The development of AKIS is attractive to the private sector. Major roads that link towns will almost certainly have to be built, or at least funded, by government. The benefits from roads will be widely spread and accrue as much to the public as to the private sector regarding access to clientele, or potential clientele. In the short and especially—the long terms, private sector issues of distribution and dynamic efficiency promise to be enhanced as a result of government commitment to AKIS.

From the Technology Transfer model to an Agricultural Knowledge and Information System (AKIS)

The 'Classical' transfer of technology model assumes that knowledge and technology are generated only through research and technological development and that these are than "transferred" mainly by an extension service to the knowledge and technology users, i.e. the farmers. Nowadays, it is recognized that information should flow in both directions. Yet, the notion of technology transfer suggest that technology can be "transferred" without change from researchers and technologists to farmers who are assume to play a rather passive role as "end receivers" in the communication chain. Furthermore, the TOT model neglects the place of policy decisions and farmers influence on them (if farmers are well organized). It also neglects the role farmers as knowledge providers and exchangers.

Genesis of Agricultural Extension

Agricultural Extension

Agricultural extension is a general term meaning the application of scientific research and new knowledge to agricultural practices through farmer education. The field of *'extension'* now encompasses a wider range of communication and learning activities organized for rural people educators from different disciplines, including agriculture, agricultural marketing, health and business studies.

To meet the challenges faced in the rural sector, agricultural research and extension have been used for decades to improve the rural economy. Agricultural extension, according to the World Bank, is "the process of helping farmers to become aware of and adopt improved technology from any source to enhance their production efficiency, income and welfare" (Purcell and Anderson, 1997). Agricultural extension has also been defined as the extending of relevant agricultural information to people (Swanson, Bentz and Sofranko, 1997). Moris (1991) calls it "the promotion of agricultural technology to meet farmers' needs". This process first became known as extension in England in 1850s (Jones and Garforth, 1997).

As mentioned by Anadajayasekeram et al. (2008), although agricultural extension has roots as far back as 1800 BC, formal extension in most countries did not start until the late 1800AD. In the United States and Canada, formal extension started during the late 1800s. France began a national service in 1879 using itinerant agriculturists; Japan and many of the British colonies also started extension service during this time. The first modern extension service was started in Ireland during the potato famine in 1845 (Swanson et al., 1997).

Agricultural extension is defined as the entire set of organizations that support people engaged in agricultural production and facilitate their efforts to solve problems; to link to markets and other players in the agricultural value chain; and to obtain information, skills and technologies to improve their livelihoods (Birner, Davis et al., 2009; Davis, 2009). This definition has evolved since the World Bank Training and Visit (T & V) program, where the focus of extension was transfer of technology to improve productivity, especially for the staple food crops. While transfer of technology still has relevance, agricultural extension must now play a wider role by developing human and social capital, enhancing skills and knowledge for production and processing, facilitating access to markets and trade, organizing farmers and producers groups and working with farmers for sustainable natural resource management (Swanson, 2008).

In short, the term 'agricultural extension', which formerly referred exclusively to public sector services, is now commonly used to include all extension-type and advisory services, as these are provided by public, private and non-governmental programs (Rivera, 2001).

Agricultural Extension System in India

The national agricultural extension system in India evolved with establishment of the Department of Agriculture in the imperial governments. Efforts to strengthen this department continued up to the time of independence. Agricultural extension was one of activities of the department and no special attention was paid to accelerate transfer of technology efforts. However, some isolated attempts were made to start special rural development programmes, including improvement of agriculture (Prasad, 1989).

Soon it was realized that sporadic and adhoc programmes might not be effective and that there was a need for sustained rural development programmes including agriculture. A nationwide, multi-purpose extension network backed with professionals became indispensible. Consequently, 55 community development projects were started in 1952. Each project covered villages with a village level worker for a group of 10 villages. For this project, extension officers-technical persons in agriculture, animal husbandry, cooperation, village industries and rural engineering were provided. The programme was based on the philosophy of integrated rural development. In 1953, the National Extension Service Programme identical to the community development programme but less resource intensity, was launched with a view to cover the entire country by 1960-61. The programme aimed to accelerate the pace of rural development, including increased employment and production by application of scientific methods in agriculture. The programme greatly emphasized the principle of development through self-help and peoples participation. The central government largely bore the cost of programme. The central government also launched several schemes to achieve self-sufficiency in food production.

The important programmes were Intensive Agricultural District Programme (1960) and Intensive Agricultural Areas Programme (1964). These programmes concentrated on the transfer of 'package of practices' and supply of critical inputs to farmers. In other words, extension strategy combined technical information with the supply of inputs. However, this strategy was discontinued with the reorganization of the extension system under the training and visit (T & V) system in 1974.

Training and Visit System of Extension

The training and visit (T & V) system, the brain child of Dr. Daniel Benor as the world bank consultant was introduced in India in 1974 for all round development of agricultural extension system in the country to introduce observation, training and technology transfer to the farmers and extension workers so as to enable them achieving greater productivity and production in the agricultural sector.

The new agricultural extension strategy was developed for accelerating economic growth and reduce absolute poverty from the poorer nations by the end of 20th century on the suggestion of Robert S. Mc. Namara, the then president of World Bank in its annual conference held at Nairobi in 1970. This was a World Bank assisted project and introduced in number of countries beginning first in Israel. The T & V system emphasized single-purpose professional extension workers, regular training of extension personnel and transfer of technology through personal contact with farmers. This concept was further strengthened through establishing research-extension-farmer linkages under the National Agricultural Extension Project (NAEP) in 1979.

The T & V system differed from earlier approaches to agricultural extension in India mainly in the responsibilities that were assigned to the front-line field officers in the villages and in the organizational structure of the state's extension services. The T & V system sought the full-time commitment of VEWs, with no responsibilities other than disseminating technology to farmers. Organizationally, the institution of a single line of command provided that the VEWs would be both technically and administratively supervised through a chain of command under the extension headquarters (The World Bank Group, 2012).

From the 1970s to the early 1990s, the Training and visit (T & V) project was implemented with the support of the World Bank. T & V was the primary machinery for technology dissemination in the public-sector agricultural extension system of India. However, it worked only through each state's department of agriculture because the focus was on crop production, particularly cereal crops. It was first introduced on a pilot basis in the Chambal command area in Rajasthan and Madhya Pradesh, with positive results. From there it was extended to the other states through the World Bank assisted National Agricultural Extension Projects: NAEP-I in Madhya Pradesh, Rajasthan and Odisha; NAEP-II in Haryana, Karnataka, Jammu and Kashmir, and Gujarat; and NAEP-III in Uttar Pradesh, Assam, Himachal Pradesh and Bihar (Glendenning and Babu, 2011).

Agriculture Extension System through Media and ICT

Due to widespread availability of ICTs such as mobile phones, internet, television, etc. digital technology has shown a tremendous potential to disseminate information to the farmers and promote extension. Several experiments of varying scale are observed across the country. MS Swaminathan Research Foundation (MSSRF) in Pondicherry has taken up the use of ICTs to disseminate information on ecology, livelihoods, socio-economic and gender aspects. Several web portals (ikisan.com, krishivihar.com, agriwatch.com and commodityindia.com) offer internet-connected farmers a variety of information and advice. The increasing growth in the use of smart phones suggests that this source of information will expand significantly in coming years. An IT app called e-sagu provides farm specific expert advice to farmers throughout the crop duration. Agencies such as IFFCO Kisan Sanchar Limited (IKSL),

Reuters Market Light (RML) and Tata m-Krishi are similar examples of knowledge dissemination to farmers through mobile phones. ICT in general did not contribute effectively in putting new knowledge into use as it mimics the traditional approach of information transfer and training of the farmers and also has limited reach (Sulaiman, 2012). None of the initiatives listed reaches out to more than a few thousand farmers, mostly in short periods. A scalable intervention leveraging ICT has yet to emerge.

Broad Observations on National Level Extension Efforts

The following broad conclusions may be drawn from the foregoing review of public, private and civil society led extension interventions in the country:

i. Agriculture extension services in India are predominantly centered around crop husbandry with a pronounced tilt towards terms of trade. The approach of public sector extension is to offer a one-size-fits-all product to all farmers. In a country with over 86% of farmers categorized as small and marginal, this is a self-limiting approach as the huge variations in resource endowment, agroclimatic conditions and legal exigencies are not factored into the model of agriculture extension being followed either by the government.

ii. While NGO-led extension models offer far more variety and display sensitivity to local priorities and conditions, they do not have the capacity or scale to make a significant impact across large regions. They are also seriously hampered in scaling up due to paucity of resources, as public sector extension agencies rarely explore synergies or cooperation and do not support continues to be project-driven and episodic.

iii. In recent years, the growth in the high value agriculture (HVA) sector has been twice or sometimes even thrice that of the crop husbandry sector. Yet agri extension services for HVA sectors remain weak and disorganized.

iv. The above analysis also suggests that the government, private sector, NGOs and others providing agriculture extension services are working in isolated silos with little or no functional coordination at the field level. This leads to restriction of good practices generated in each of these sectors and an opportunity for wider application is lost.

v. Lastly, it may be concluded that the large number of players in the agriculture extension arena function without any standards or certification of quality. This leaves questions of accountability up in the air as the majorities of farmers are not in a position to pursue legal remedies in case of erroneous or even harmful advice.

Salient recommendations of Sub-groups on Agricultural Extension

• Extension is the most important instrument for agricultural development, hence, extension Reforms Scheme may be up-scaled in all the 588 development districts in the country for strengthening extension system by establishing agricultural technology management agencies (ATMAs).

• There is a need to establish a well equipped State Agricultural Management and Extension Training Institute (SAMETI) at state level, full complement of staff in ATMA, and strengthening a training institution at the district level for effective delivery of extension system.

• Since all the districts have Krishi Vigyan Kendras (KVKs), generally one per district, the mandate and functioning of KVKs should be clearly defined, and KVKs and ATMAs must complement activities of each other and should avoid duplication. Synergy between these institutions should be clearly promoted.

• Progressive farmers from different commodity groups may be identified, trained at District, State and National level. They may be recognized as resource persons in extension activities, further promoting the farmer-to-farmer extension.

- Wide publicity is required for popularizing Kisan Call Centres among farming community and other stakeholders.
- Diploma in Agricultural Extension Services for Input Dealers (DAESI) programme may be up scaled to entire country so that the services of input dealers can be utilized as Para extension workers at village level.
- Create mechanism at district and block levels for integration of the efforts of multiple agencies involved in research and extension activities both in public and private sectors.
- To monitor effectively the Operationalisation of Research—Extension linkages, a separate Monitoring and Evaluation Cell (Research-Extension-Farmer and Market Coordination Committee) to be operationalised with the Agricultural Production Commissioner (APC) in each state. Similarly, at district level ATMA should ensure R-E-F-M linkages.
- Farm schools and farmers organizations would be promoted on a massive scale in extension work.
- National Institute of Agricultural Extension Management (MANAGE) may be promoted as International Centre of excellence in Agricultural Extension Management laying greater emphasis on capacity building activity at National level and SAMETI at State level.
- Establishment of a national forum for research and extension as a clearing house for various developmental issues in research and extension.
- Adequate provisions are made under Extension Reforms for involvement of PRI's, which need to be effectively implemented.
- Agri-preneurs may be identified, trained under Agri-Clinics and Agri-Business Centres Scheme and established with the help of ATMA's. Such Agripreneurs may be linked to commodity specific groups. AC-ABCs may also be associated in implementation of ATMA activities.
- Agri-Clinics and Agri-Business Centres Scheme may be continued with strengthened landholding mechanism involving state governments, ATMA's, Agri-Business companies and Banks.
- Agro-Polyclinics may be established at block level by private sector or in Public Private Partnership (PPP) mode to provide multiple services to farmers.
- Awards may be instituted at different levels in recognition of services/achievement of farmers and extension service providers.
- Sustained Capacity Building of Public Extension Functionaries should be taken up for providing quality extension services.
- Extension intervention by the private sector should preferably be in project mode with quantifiable results and accountability aiming at maximizing profit of farmers. Private initiatives may be supported by government schemes, fiscal incentives.
- There is a need to strengthen Extension Education Institutes (EEIs) as Zonal level Training Institutions liberating from State Agricultural Universities (SAUs) control.
- In all the trainings programmes at all levels, women participation must be ensured. Specialized training for women trainers; women farmers have to be organized. Across all the programmes, allocation of minimum 30% of the resources may be ensured for training of women functionaries/women farmers.

- Involvement of private training/Human Resource Development/consultancy institutions in training programmes in specialized areas.
- Placement of core faculty in management areas is needed at SAMETIs to inculcate managerial competencies in extension functionaries.
- To organize the Farmers field schools on massive scale, adequate trained manpower may be ensured by State Agricultural Universities, KVKs, SAMETIs and State Agriculture and allied sector departments.
- Training Need Assessment (TNA) at grass root level to be conducted by ATMAs/ SAMETIs.
- Specific schemes like Women in Agriculture with modifications should be continued till gender mainstreaming is made operational at various levels.
- All the stakeholders in agricultural extension may be sensitized in gender issues.
- Self Help Groups may be supported with better credit, marketing facilities and support services.
- Gender budgeting may be introduced meaningfully in agricultural extension programmes.
- The programme content, service delivery and the business models of Mass Media and Information and Communication Technology (MM & ICT) should specially address the needs of small and marginal farmers.
- KVKs should play a key role in support of the Mass Media and ICT initiatives in each district. The Community Radio Centres may be established at each KVK so as to provide location specific information to the farming community.
- A dedicated TV channel for agriculture may be launched, if feasible.
- Mass Media and ICT attempts by public, private and NGOs in the field of agriculture and related sectors need to be converged and shared among them.
- The gaps identified in Strategic Research and Extension Plans (SREPs) of each ATMA district and Frequently Asked Questions (FAQs) of Kisan Call Centers may be used for production of the messages/programmes for Radio, TV, print and content for training Programmes.
- There is a need to improve the awareness among farmers on Kisan Call Centres (KCC)—particularly its cost free services through toll free telephone so as benefit needy farmers.
- Expertise of level-I of KCC needs to be improved.
- FM Radio should have a provision for advertisement while broadcasting of agriculture programmes as a business model to generate revenue under Public-Private-Partnership mode.
- For decentralized decision-making, effective coordination, monitoring and concurrent evaluation of centrally sponsored programmes, Zonal level coordination units similar to Zonal coordination unit of KVKs need to be established.

Source: (Recommendation of Working Group on agricultural extension, Planning Commission, Government of India, 2007).

Participatory
Extension Approaches

Animation Rural

Animation rural was the first systematic attempt to introduce participatory methods into extension systems. This approach, introduced by the French in francophone Africa, was based on a participatory, emancipator philosophy with parallels to the philosophy of Paulo Freire in Brazil (Nagel, 1997). The approach helped raise group consciousness and collective action to define, understand, and address local problems and to integrate rural areas into national systems and programs. A primary feature was the *animateur* or *animatrice*, an individual not already involved in village leadership and selected by the village to be trained, supervised, and supported by the government's rural development agency. The trained individual would reside in the village, share his or her knowledge and skills with other villagers, and serve as a communication link between the village and government agencies.

This approach was not easy to operationalize and has not been formally continued in most countries. Farmers often wanted technical information more than just "consciousness raising." However, this approach to empowerment, consciousness raising, and participatory development is still in use today, particularly among nongovernmental organizations (NGOs).

Integrated Rural Development

Starting in the 1960s and continuing until the present time, there have been various attempts to pursue a more holistic, community or integrated approach to rural development. For example, Yudelman (1976) pointed out that during the 1960–70s, rural development projects funded by the World Bank focused on increasing the output and incomes of low-income producers, often by means of the introduction and expansion of technological change at the farm level. The assumption underlying this effort is that three basic conditions must be met if changes are to be brought about. First, producers must know how to increase their output; second, they must have access to the means of increasing their output; and third, they must have the incentive to make the effort and accept the risk associated with increasing their output (Yudelman, 1976). It is interesting to note that these three conditions are just as applicable today as in 1976 and should be given full and careful consideration in determining ways of improving extension systems to improve rural livelihoods.

During the 1970s and 1980s, these approaches largely subscribed to a "one size-fits-all," top-down approach being implemented through the T & V extension model. In the late 1980s and 1990s, it became clear that a more inclusive and targeted rural development approach was needed so that the rural poor could be empowered to spur development. The importance of local ownership was recognized, as was the effectiveness of a more people-centered, multi stakeholder approach.

The role of public agricultural extension in these emerging integrated community and rural development programs was limited. Rural development departments that used a more multisectoral approach generally implemented broader community development programs. Learning from the past, the UN Economic and Social Council (2003) has indicated that new approaches to integrated rural development should be based on a territorial (i.e. community), rather than a sectoral logic, emphasizing location-specific synergies both within and among different sectors.

In conclusion, the objective of most donors in reducing rural poverty is to help countries accelerate economic growth so that the rural poor can share the growth benefits. This strategy focuses on (1) fostering an enabling environment for broad-based and sustainable rural growth; (2) enhancing agricultural productivity and competitiveness; (3) fostering nonfarm economic growth; (4) improving social well-being, managing and mitigating risk, and reducing vulnerability; and (5) enhancing the sustainability of natural resource management.

Key Constraints in Creating a More Integrated Rural Development System

Since the 1980s, there has been a growing recognition of the need to engage different groups of farmers in setting research and extension priorities. For example, rapid rural appraisal (RRA) techniques were developed in the 1970s and 1980s in response to the perceived problems of researchers, who lacked good connections with local people in identifying important farming system constraints. The problem is that many of these farming system issues are very location specific, due to differing agro-ecological conditions, farmer needs, and access to markets. Therefore, research and extension activities concerned with the diversification of farming systems are more difficult to scale up, than merely transferring technical information about new wheat or rice varieties to farmers.

Subsequently, participatory rural appraisal (PRA) tools evolved from the RRA methodology into a new set of techniques that could be used by rural development practitioners and field extension workers to collect and analyze data on local problems, including socioeconomic factors (World Bank 1996a). Conducting PRAs was part of a growing family of participatory approaches and methods that emphasized the importance of local knowledge in program planning and in enabling local people to carry out their own needs assessment as they help shape extension plans and priorities. In short, the key tenets of PRAs include participation, teamwork, flexibility, and triangulation.

The difficulties in moving from a *top-down* extension system to one that is more *bottom-up* is tied directly to shifting program planning from national/provincial extension directors to the clientele being served at the district and sub-district levels. However, merely having extension workers conduct PRAs will do little good unless representative farmer groups have a formal framework (i.e. advisory or steering committees) through which they can articulate their needs and help set research and extension priorities for different groups of farmers at the district and sub-district levels. For example, in transforming the Indian extension system, *governing boards* were established at the district level both to review extension programs and then to allocate resources to extension offices at the sub-district level. In addition, *farmer advisory committees* were established at the sub-district level, including women farmers and other disadvantaged groups, so these different clientele groups could both articulate their priorities and needs, as well as assess the performance of extension workers who are delivering these needed programs.

In short, if the management structure of extension systems is not properly organized, then the needs of larger, commercial male farmers will probably take priority. On the other hand, if extension systems are going to improve rural livelihoods, especially among the rural poor, then the district and sub-district extension offices must fully engage small-scale men and women farmers in both setting priorities (including in research) and in delivering needed programs. In many cases, these opportunities will focus on labor intensive, high-value crops, livestock, and other enterprises that can substantially increase farm income and thereby improve the livelihoods of the rural poor. However, most government agencies, including extension, are unwilling to establish these formal participatory mechanisms unless forced to do so by national policy makers and/or donor agencies that recognize the long-term benefits of *formal* stakeholder participation in shaping extension programs and priorities.

Farmer-Based Extension Organisations

The best example of a fully demand-driven extension system is one that is directed, operated, and financed by farmers themselves. Depending on the country, these extension systems generally operate under different management structures and with different sources of financial support. It is important to note that large-scale, commercial farmers who have better leadership and better organizational and technical skills, as well as more economic power, frequently dominate these farmer-controlled extension systems. Therefore, bringing the rural poor (including women farmers and other disadvantaged groups) into these systems will take considerable time and effort, especially in developing their leadership and organizational skills.

Most farmer-operated extension systems are found in industrially developed countries where commercial farmers have attained the organizational capacity to effectively manage these service agencies. For example, the Danish Agricultural Advisory Service (DAAS) has over 3,600 staff members and is solely under the direction of farmer organizations. The average contribution of each farmer to DAAS is about $10,000/year. The Agricultural Advisory System operated by *Chambres d'agriculture* in France has over 7,000 technical staff and continues to provide useful services to all groups of farmers within France. This system is financed by a mandatory land tax on the size of each farm.

Successful examples of farmer-controlled extension systems are beginning to emerge as well in developing countries, such as in Chile). In that country, extension services have been gradually privatized over the past 30+ years, but these systems are still publicly funded. Private-sector firms deliver advisory services to each of the participating farmer organizations, but they primarily provide the following types of extension services: farm management, post-harvest handling, value-added processing, and legal services. Generally, they do not focus on the traditional technical advisory services designed solely to increase agricultural productivity.

MARKET-ORIENTED EXTENSION APPROACHES

Commodity-Based Advisory Systems

Advisory services for major export crops have been in existence since colonial times and are still common in many developing countries that produce major export crops, such as rubber, tobacco, coffee, cocoa, sugar cane, oil palm, bananas, oranges, and cotton. Generally, a private-sector firm or a parastatal organization is responsible for operating these commodity-based advisory systems. These advisory systems are generally both effective and efficient because they generally serve specific agro-ecological areas where these export crops can be grown and the advisory personnel work solely with those contract farmers who are growing these particular crops. Because these advisory services are limited to just one commodity, training of both the advisory agents and farmers they serve is relatively simple and straightforward. In addition, the farmers themselves have an economic interest in following these recommended practices so they can sell their respective crops.

In most cases, financing of both research and advisory services for these export crops is generated by a "cess," or tax, which is paid for by the participating farmers, based on the quantity and value of products being sold to exporters. Generally, this tax is about 1% or less of the gross income paid to these farmers. In summary, most export commodity-based advisory systems are well organized and financed; therefore, they are both effective and efficient in providing these advisory services to the participating farmers. For an example of a commodity-based (cotton) advisory system in Mali (Bingen and Dembèlé, 2004). Regardless of which management model is used, it is critical that the primary stakeholder groups be formally involved in setting research and extension priorities as well as in assessing how these program funds should be used.

In addition to these export commodity-based advisory systems, there are other excellent examples where advisory and other services are being provided to producer group members by either their cooperative or by private-sector export firms. For example, the Gujarat Cooperative Milk Marketing Federation, which has now been operational for 35 years, currently has about 2.8 million members across India who daily delivers milk through over 13,000 village societies. In addition to these village societies collecting and processing about 8.4 million litres/day, they also provided artificial insemination (AI) services to over 3.5 million cows owned by its members (Gujarat 2009). Another excellent private-sector example is HJS Condiments in Sri Lanka, which provides advisory services to about 8,000 of its farmers who produce export crops, such as gherkins or pickles. The agent-farmer ratio is about 1 to 100, and these advisory services are fully financed by the company itself. The field advisors make about one visit every two weeks during the growing season and primarily focus on production practices, quality control, and postharvest handling procedures. This company is continuing to expand its export of horticultural products, and the number of participating farmers continues to increase.

Innovative, Market-Driven Extension Approaches

The emerging market-driven model of organizing extension systems is a 180° change in direction from the traditional linear model of linking research to extension to farmers, illustrated earlier by the technology transfer model, to an emerging new innovative extension model. This innovative, market-driven approach is consistent with the agricultural innovation systems framework, especially within a rapidly changing global economy. In short, where there is economic development, there are generally changes in consumption patterns that create emerging markets for new high-value products (Step 1).

Under this emerging new extension approach, it is the growing market for high-value products—not research—that controls specific innovations that can be successfully taken up by different farm households within local communities to improve their farm household income. In the process, each farm household must consider its own resources (e.g. land, labor, access to water) and access to different markets (e.g. transportation infrastructure; distance to different local, regional, and even global markets). Then, it must determine which enterprises would be most feasible and whether appropriate technologies are easily available for them to successfully produce and market these different crops, livestock, fisheries, or other agricultural products.

Other critical factors that influence the success of this market-driven extension approach include these:

- Helping farmers, including farm women, who have similar resources and interest; organize into producer groups within each community.
- Having access to reliable markets and market information about which crops and products have sufficient economic potential to be produced and marketed in the different agro-ecological zones within each district.
- Having access to production inputs (e.g. seeds) and training in the production and other management practices necessary to successfully produce these different high-value crops or products and to meet market specifications. It should be noted that urban consumers are increasingly influenced by global food preferences. Therefore, it may be necessary to secure new varieties from international sources (e.g. private-sector companies), because these planting materials may not be readily available locally.
- Product identification and certification process (knowledge and access).

It is important to remember that a market-driven extension approach helps farmers move incrementally toward agricultural diversification. Small-scale subsistence farm households do not stop producing the basic food crops needed for home consumption. Rather, they allocate a small amount of their land to produce a specific high-value crop (e.g. fruits or vegetables) or product (backyard poultry, honey, mushrooms, etc.) and, after they work out the necessary production and marketing practices. Then, they begin scaling up the production of these crops or products, based largely on profitability.

In most cases, male farmers focus on those field or export crops that are more in line with cultural tradition, while women farmers generally pursue a different set of high-value crops or products that are traditionally grown by women in that particular culture. In many countries, women may be better positioned to pursue this market-driven approach, due to cultural traditions and their labor availability during different parts of the day. For example, in some cultures, women may be better suited to undertake enterprises that are closer to their home, such as backyard gardening and poultry, caring for and milking dairy cows, producing mushrooms, and vermicomposting.

Briefly, the steps involved in this process start by having district and sub district extension workers use PRA techniques to first identify innovative farmers within their district who are producing different types of high-value crops or products, or who are using innovative production techniques (Step 2). Then, the district extension subject matter specialists (SMSs) need to assess these innovations (within the district) with researchers, to determine whether these innovative farmers are using the most up-to-date production methods and materials. In addition, the district extension office has to do some market research to determine the potential demand for these products, both locally, regionally, nationally and, possibly, globally. If there is potential for expanding the production of specific products, then potential areas within the district can be identified. The criteria used will involve both agro-ecological conditions and access to markets, which will determine the relative comparative advantage of different communities in producing and marketing specific crops or products. At this point, the field extension staff will have to begin presenting and discussing these different options with both men and women farmers within each community to assess their possible interest in pursuing one or more of these different market opportunities.

The next steps are to take interested farm leaders from different communities to visit with and discuss these potential enterprises with these innovative farmers (Step 3). Most men and women farmers are open to listening to a progressive farmer before they will fully trust a new idea from a local extension worker. However, once they are convinced by listening to an innovative farmer, then they will be ready to ask the extension staff for help in learning how to produce and market these new products. At this point, the district extension office must work with research to obtain the necessary technical and marketing information and/or to engage researchers in actually training the first group(s) of interested farmers (Step 4). At the same time, the local extension staff need to work with all interested farmers within each community to begin organizing producer groups, first at the community level and, subsequently, linking these community groups together as *producer associations* within the district (Step 5). Finally, as these groups get started with the first production season, then extension will have to assist these groups in working out the post-harvest handling and marketing of these products (Step 6).

NON-FORMAL EDUCATION/EXTENSION APPROACHES

Farmer Field Schools

The farmer field school (FFS) approach to organizing extension programs began in Indonesia over two decades ago as a means of educating farmers how to incorporate integrated pest management (IPM) practices into their farming systems, especially for rice production. This approach primarily uses non-formal education methods to teach farm leaders in each community how to reduce pesticide use, which in turn helps increase farm income. Based on an impact evaluation of 25 different case studies,

Van den Berg (2004) concluded that farmer field schools had a significant impact on reducing the use of pesticides and increasing yields. Perhaps more importantly, however, this approach stimulated continued learning and strengthened the social and political skills of farmers. In some countries, these developments triggered a range of local development activities, relationships, and policies. As the FFS model has been implemented in Sub-Saharan Africa, this non-formal education approach has been expanded to cover an increasing range of production practices, most with an individual crop production focus (Davis et al, 2009).

University-Based Extension

The US Cooperative Extension Service model originated with and continues to be managed by land grant universities in each state. Federal, state, and local (county) governments jointly fund this extension system. Most state extension systems focus on four primary areas:

a. Agriculture and natural resource management

b. Consumer sciences, including family nutrition, health, and financial management

c. 4-H and youth programs

d. Community and economic development

This extension system continues to emphasize non-formal education activities in each of these major program areas, which may differ somewhat from state to state.

This decentralized extension system has an extension office in nearly every county within each state. Most important is that primarily the local county extension advisory committee, with most program funds and transportation costs being provided by the local county government, determines program priorities. Most subject-matter specialists are located at the land grant universities in each state, and most have joint research and extension appointments, so there are strong linkages between research and extension at the state level. These subject-matter specialists provide regular training programs for extension educators as well as certified crop (technical) advisors from private-sector firms. In fact, all private-sector agricultural advisors are required to complete 40 hours of professional training every two years to remain certified. This approach ensures that farmers receive up-to-date and accurate advisory services from both public extension (mainly through the Internet and conferences) and private-sector advisors (one-on-one advisory services to clients). Federal and state governments provide joint funding for state-level extension operations (primarily salaries, research support, and operating funds), while county governments provide funding for most local extension program activities. This administrative and financing structure enables the Cooperative Extension Service system to adapt to the changing economic, technical, and social developments within each state and county.

Concept of Farm School

1. The National Commission on Farmers has recommended that Farm Schools may be established in the fields of outstanding farmers. Such farm school will be based on the principles of "Learning by doing" as well as 'seeing is believing' with focus on farmer-to-farmer extension. The farm school would help in developing a cost effective extension systems.

2. The key features of the Farm Schools to be promoted under the ATMA programme are given below:
   - Farm Schools would be operationalized at Block/Gram Panchayat level.
   - These would be set up in the field of outstanding farmers and awardees of nationally recognized awards for farmers. These could also be set up in a Government/Non-Government institution.
   - "Teachers" in the Farm Schools could be progressive farmers, extension functionaries or expert belonging to Government or Non-Government sector.
   - One of the main activities of Farm School would be to operationalize Front Line Demonstrations in one or more crops and/or allied sector activities. These demonstrations would focus on Integrated Crop Management including field preparation, seed treatment, IPM, INM, etc.
   - Farm Schools would provide season long technical back stopping/training to target farmers.
   - The "students" of Farm Schools as per specified schedule or as may be necessary. "Teacher" may also visit students as may be necessary.
   - Knowledge and skills of "teachers" would be upgrades on a continuous basis through training at district/state/national level institutions and exposure visits, etc.
   - In addition to technical support through Farm Schools, knowledge and skill of "students" may also be upgraded through training at district/state/national level institutions and exposure visits, etc.
   - "Students" would have the responsibility of providing extension support to other farmers in the respective village or neighboring villages.

National Agricultural Extension Project (NAEP)

The basic objective of NAEP was to bridge the gap between research system with that of extension system so that the transfer of technology takes place at a much faster rate resulting in higher production and prosperity in the rural sector in general and agricultural sector in particular.

Four major paradigms of agricultural extension: The terms extension and advisory services can be used somewhat interchangeably, but the following framework gives a useful perspective on the different approaches being pursued by different countries and donors in organizing and implementing effective extension systems. This framework juxtaposes these different terms or approaches by reviewing how the delivery of educational programs and information/communication services takes place and why it takes place. In this framework, the options are whether extension workers want to convince farmers what to do (i.e. persuasive methods) or whether they seek to inform and educate farmers about different market opportunities, technical options, and/or management strategies, and then let them decide which option would work best for them. The following classifications illustrate different combinations that help describe and highlight important differences between these different approaches or paradigms in organizing agricultural extension and advisory services (Swanson 2008b, p. 6)

- Technology transfer—this extension model was prevalent during colonial times and reemerged with intensity during the 1970s and 1980s when the Training and

Visit (T & V) system was established in many Asian and Sub-Saharan African countries. This "top-down" model primarily delivers specific recommendations from research, especially for the staple food crops, to all types of farmers (large, medium, and small). This approach generally uses persuasive methods for telling farmers which varieties and production practices they should use to increase their agricultural productivity and thereby maintain national food security for both the rural and urban populations in the country. The primary goal of this extension model is to increase food production, which helps reduce food costs. As illustrated by North American and European countries, as farming becomes increasingly commercialized, both technology development and transfer will increasingly be privatized.

- Advisory services—both public extension workers and private-sector firms, in responding to specific farmer inquiries about particular production problems, still commonly use the term advisory services. In most cases, farmers are "advised" to use a specific practice or technology to solve an identified problem or production constraint. Public extension organizations should have validated information available from research about the effectiveness of different inputs or methods in solving specific problems so that inquiring farmers receive objective and validated information. Most input supply firms use persuasive advisory techniques when recommending specific technical inputs to farmers who want to solve a particular problem and/or maintain their productivity. Although most firms use persuasive methods to sell more products and increase their profit, an alternative private-sector model is to support out grower schemes where export firms have field agents who both advise and supervise contract growers to ensure that specific production inputs and practices are followed.

- Non-formal education (NFE)—in earlier days of extension in Europe and North America, this paradigm dominated when universities gave training to rural people who could not afford or did not have access to formal training in different types of vocational and technical agriculture training. This approach continues to be used in most extension systems, but the focus is shifting more toward training farmers how to utilize specific management skills and/or technical knowledge to increase their production efficiency or to utilize specific management practices, such as integrated pest management (IPM), as taught through farmer field schools (FFS). Both NFE and facilitation extension (as described next) commonly help farmers with similar resources and interests to organize into different types of producer or self-help groups, particularly if they want to learn how to diversify or intensify their farming systems, especially in pursuing new, high-value crops or other products.

- Facilitation extension—this approach has evolved over time from participatory extension methods used 20–30 years ago and now focuses on getting farmers with common interests to work more closely together to achieve both individual and common objectives. An important difference is that front-line extension agents primarily work as "knowledge brokers" in facilitating the teaching–learning process among all types of farmers (including women) and rural young people. Under this extension model, the field staff first works with different groups of farmers (e.g. small-scale men and women farmers, landless farmers, etc.) to first identify their specific needs and interests. Once their specific needs and interests have been determined, then the next step is to identify the best sources of expertise (e.g. innovative farmers who are already producing and marketing specific products, subject matter specialists, researchers, private-sector technicians, rural bank representatives) that can help these different groups, address specific issues and/ or opportunities. Agriculture and rural development innovative farmers have already worked out the necessary practices to successfully produce and market these new crops and/or products. In short, innovative farmers are frequently the starting point for extension workers who want to facilitate the intensification and diversification of farming systems to increase farm household income. In many cases, these innovative farmers, if properly approached, can be encouraged to become the leaders of these new producer groups, which will both enhance their reputation within the community as well as increase profits for all members by expanding their supply of high-value products to larger urban markets. Once other farmers become interested in pursuing specific new market opportunities, then both research and extension will need to work in close collaboration with these innovative farmers in advising the "start-up" farmers on the most applicable practices and technologies. In the process, these front-line extension staff will have to facilitate the training and backstopping of these farmers during the first year or two in producing these new crops, livestock, or other enterprises. When small-scale farmers become interested in pursuing these types of new economic opportunities, they are ready to engage in an active learning process. This innovative, market-driven extension approach works best where men and/or women farmers are already interested in intensifying and/or diversifying their respective farming systems with the goal of increasing farm household income. This facilitation approach can also be used to train members of landless households, especially rural women, how they may be able to use common property resources (CPR) to start new enterprises and thereby increase their household income. As will be discussed in the next section, all four of these extension models or paradigms have an important role to play in helping achieve different agricultural development objectives. However, to both increase farm income and improve rural livelihoods among the rural poor, it will be necessary for most public extension organizations to transition toward greater use of facilitatory and NFE extension methods. In particular, small-scale men and women farmers, including the landless, can begin organizing into community or farmer groups and then learn the necessary technical, management, and marketing skills that will be necessary to help them progressively diversify into higher-value crop, livestock, or other enterprises that will increase their farm household income. At the same time, as the agricultural sector in countries develops (i.e. becomes increasingly commercialized), technology transfer and advisory services tend to be increasingly privatized. Therefore, in the process, it is important to build strong public–private partnerships that will further enhance agricultural productivity growth, as well as to increase the incomes and improve the livelihoods of small-scale and landless farm households. Another important change is the shift from a more linear technology transfer model toward a more holistic approach in understanding how and where farmers get their information and technologies. For example, the current move toward an agricultural innovations systems approach arises through an interactive, inclusive process relying on multiple sources and actors (World Bank, 2006b). Especially when the goal is to intensify and diversify farming systems, both innovative farmers and extension can play a significant, joint role in working together to introduce new market-driven crop and/or livestock systems to small-scale men and women farmers. Therefore, extension, in effect, serves as a facilitator or knowledge broker; this transition has implications for the technical, professional, and entrepreneurial skills that extension agents will need to be effective in this new role (Rajalahti, Janssen, and Pehu 2008).

Policy Framework for Agricultural Extension

The need for reforms in Agricultural Extension has been explicitly raised in the National Agriculture Policy; the report of Expenditure Reforms Committee, as well as, the Tenth Plan Approach paper. Keeping the recommendations of these policy initiatives in view, and to provide policy directives for extension reforms, a broad Policy Framework of Agricultural Extension (PFAE) has been developed by the Ministry of Agriculture.

The five major guiding elements of the Policy Framework are as follows:
1. Reforming Public Sector Extension
2. Promoting private sector to effectively complement, supplement and whenever possible to substitute public extension
3. Augmenting Media and Information Technology Support for Extension
4. Mainstreaming Gender concerns in Extension
5. Capacity building/skill up-gradation of farmers and extension functionaries.

PILOT TESTING OF THE REFORMS UNDER THE WORLD BANK FUNDED NATIONAL AGRICULTURE TECHNOLOGY PROJECT (NATP)

The reform enlisted above has been pilot tested under Innovations in Technology Dissemination (ITD) component of World Bank funded National Agricultural Technology Project (NATP) with effect from November, 1998 in seven states, viz. Andhra Pradesh, Bihar, Himachal Pradesh, Jharkhand, Maharashtra, Odisha and Punjab covering 4 districts in each state. An autonomous institution—Agricultural Technology Management Agency (ATMA) has been established in project districts as a registered society representing various stakeholders, including farmers, in project planning and implementation. Monitoring and Evaluation (M & E) reports of ITD component brought out by Indian Institute of Management (IIM), Lucknow reveals that the ATMAs' extension approaches have been proving to be very promising in execution of the reforms envisaged in the policy Framework for Agricultural Extension.

New Scheme of Extension Division of the DAC

The 10th Five Year Plan Approach Paper called for radical overhaul of extension services and significant improvements in sophistication of technology dissemination methodologies. It highlighted the need for specific measures to ensure that research technology development and extension services meet the special needs of farmers.

The PFAE and experiences under ATMA approach have also been directed towards similar strategies. Accordingly, the Department of Agriculture & Cooperation, Ministry of Agriculture, Government of India is implementing the following schemes during the Xth Plan period.

1. Mass media support of Agriculture Extension
2. Support to State Extension Programmes for extension reforms
3. Agri-clinics/Agri-Business centres
4. Establishment of Kisan Call Centres
5. Extension support to Central Institutions
6. Externally Aided Projects.

The Scheme: Support to State Extension Programmes for Extension Reforms

The scheme "Support to State Extension Programmes for Extension Reforms" is the main scheme to operationalized agricultural extension reforms across the country. Under the scheme funding support shall be provided to the States/UTs for undertaking extension reforms within the broad purview of the Policy Framework for Agriculture Extension (PFAE), complying with its key areas/norms, and shall be operated based on Extension Work Plans prepared by them.

Inter-alia, the following key reforms, in line with the PFAE are being promoted under this scheme:

- *New Institutional Arrangement:* Providing innovative restructures autonomous bodies at the district/block level, which are flexible, promoted bottom up and participatory approaches, are farmer driven and facilitate public-private partnership.
- Convergence of line departments': Programmes and operating on gap filling mode by formulating Strategic Research and Extension Plan (SREP) and Annual Work Plans.
- Encouraging multi agency extension strategies: Involving inter-alia public/private extension service providers.
- Moving towards integrated, broad-based extension delivery in line with farming system approach.
- Adopting group approach to extension: Operating through Farmers Interest Groups (FIGs) and Self Help Groups (SHGs).
- Addressing Gender concerns (Mobilizing farm women into groups, capacity building, etc.)
- Moving towards sustainability of extension services (e.g. through beneficiary contribution).

Project Implementation

Under the scheme funding shall be released to the States based on their Extension Work plans developed within the broad framework of the PFAE and areas indicated under the cafeteria of reform oriented activities. The States shall propose the new institutional arrangements, similar to ATMA, they intend to put in place in the first work plan, or even earlier, to be submitted by them to the DAC for approval. This agency will have the responsibility of implementing the extension reforms at districts level.

New Institutional Arrangement

ATMA would be supported by a Governing Board and a Management Committee. Under each of the ATMA, block level Farm Information and Advisory Centres (FIACs) have been created which are operated by a Block Technology Team (BTT) of technical advisors and a Farmer Advisory Committee (FAC), a group of farmers. Commodity oriented Farmer Interest Groups (FIGs) are promoted at block/village level to make the technology generation/dissemination farmer driven and farmer accountable. In order to provide needed HRD support in innovative areas of extension delivery, a State Agricultural Management and Extension Training Institute (SAMETI) has also been established in the project states.

STATE LEVEL: INTERDEPARTMENTAL WORKING GROUP (IDWG)

In pursuance of the number of mechanisms built into the project design and to ensure effective coordination among the departments like agriculture, animal husbandry, fisheries, horticulture, soil conservation, etc. it is proposed to constitute a state level inter departmental working group under the chairmanship of the Agriculture Production Commissioner/Secretary Agriculture with the following composition.

Composition

In departments like horticulture, soil conservation, etc. where separate secretaries do not exist, director of the concerned departments may act as a member on the interdepartmental group.

Key Function of IDWG

- To provide mechanism for interaction with the Technology Dissemination Management Committee (TDMC) of the DAC, GOI, guide the human resource development activity and to monitor the district level technology dissemination programme.
- To oversee and support Agricultural Extension Research activities being undertaken by ATMA and to make policy interventions or interdepartmental matters including issues related to Women in Agriculture and Coordination thereof.

Table 8.1: Composition of Interdepartmental Working Group (IDWG)

S. no.	Member	Designation
1.	Agriculture Production Commissioner/ Secretary Agriculture	Chairman
2.	Secretary (Finance)	Member
3.	Secretary (Fisheries)	Member
4.	Secretary (Horticulture)	Member
5.	Secretary (Rural Development)	Member
6.	Secretary (Animal Husbandry)	Member
7.	Secretary (Soil Conservation)	Member
8.	Secretary (Women Development)	Member
9.	Secretaries of related departments (wherever necessary)	Member
10.	Vice-Chancellor(s) of SAUs	Member
11.	Secretary (Agri)/ Deputy Secretary (Agri.)	Member

- To promote and establish integrated approach in transfer of technology at State, division and district level by agriculture and line departments.
- To establish effective linkages with different line departments, marketing, input and credit institutions, NGOs, Private/corporate sector to promote large scale extension reforms.
- To internalize new concepts and institutional arrangement successfully demonstrated by the ATMAs.
- To deal with any other policy issue related to implementation of the project, which emerges from time to time.

A **State Nodal Cell (SNC)** would be established in each States with the office of the Director of Agriculture. This SNC would monitor project activities being carried out in each pilot district and ensure that project funds released to the States are included within state budget.

STATE AGRICULTURAL MANAGEMENT AND EXTENSION TRAINING INSTITUTE (SAMETI)

The State Agricultural Management and Training Institutes (SAMETI) would be strengthen by way of providing training, managerial and equipment, communication support, programme cost SAMETIs are proposed to be autonomous institutes with greater flexibility in structural and operational aspects.

Key Functions of SAMETI

- To provide capacity building support in Extension Management related areas to the extension functionaries both from public and private sector.
- To develop consultancy in the areas like project planning, appraisal, implementation, etc.
- Develop and promote the application of management tools for improving the effectiveness of Agricultural Extension services through better management of human and material resources.
- Organize need based training programs for middle level and grass-root level agricultural extension functionaries.
- Develop modules on management, communication, participatory methodologies, etc. as a sequel to the feedback from training programs.

FARM INFORMATION AND ADVISORY CENTRE (FIAC)

It consists of two bodies namely Farmer Advisory Committee (FAC) and Block Technology Team (BTT). The FAC is a body of farmer representatives (11–17) members representing various enterprises and socioeconomic strata. The BTT on the other hand is a group of technical advisors operating at block level representing agriculture and allied sectors. FAC and BTT took together, act as planning and operational arm of ATMA. It would be created at the block level. BTT convener will act as the convener of FIAC also.

It would, in effect, manage key extension programmes within the block level, leaving other service and developmental activities to be managed by other units or personnel within the line departments.

It would be the common meeting point for line departments to prepared detailed extension programmes and coordinate their implementation. It would also be the level where farmer input could be more effectively mobilized through a Farmer Advisory Committee (FAC). Such a mechanism, including representatives of all major stakeholders, would help set extension priorities across each program area and allocate resources.

The FIAC team would be responsible for operationalizing the SREP in each block and moving toward a single window extension system. The FIAC team would prepare Block Action Plans (BAPs) that would detail extension activities to be undertaken. This plan should be approved by the FAC, before it could be forwarded to the ATMA. The ATMA Management Committee (MC) would ensure that these plans were technically and administratively feasible, and consistent with the SREP, before being forwarded to the ATMA Governing Board (GB) for approval. The district–level line departments and research units would also prepare annual WPs to (1) maintain diagnostic and support services (e.g. soil testing laboratories), (2) organize in service training and technical support activities for FIAC and field level extension staff, (3) carryout research programs and (4) periodically update the district SREP.

BLOCK TECHNOLOGY TEAM

It is an Interdepartmental Team of Agriculture and Line Departments operating at block level. An indicative composition of BTT is given below; however the composition would change from place to place depending on critical areas pertaining to different states.

Composition

Block level officers of Agriculture, Horticulture, Animal Husbandry, Fisheries, Veterinary Science, Soil Conservation, Sericulture, Cooperative, Marketing, etc.

In West Bengal one key grass root level extension workers (like KPS, etc.) of Agriculture and another one from any line department are included in BTT.

The Agricultural Development Officer heads the Block Technology Team as BTT Convenor.

Key Functions of Block Technology Team (BTT)

The key functions of Block Technology Team would be to:
- Operationalized the SREP in each block and move towards single window extension system
- Help district core team in upgradation of SREP
- Prepare block action plan detailing extension activities to be undertaken
- Coordinate the implementation of extension programs detailed in the Block Action Plan
- Facilitate formation of FIGs/FOs at the block level and below.

FARMER ADVISORY COMMITTEE

The Farmer Advisory Committee consists of 11–17 members covering different categories of farmers covering under the given block, with due representation to

Table 8.2: Composition of Farmer Advisory Committee

S. no.	Member	Occupation
1.	Farmer	Agriculture
2.	Farm Women	Agriculture (SC)
3.	Farmer	Horticulture
4.	Farm Women	Horticulture
5.	Farmer	Livestock Producer
6.	Farm Women	Livestock Producer (SC)
7.	Farmer	Fisheries
8.	Farm Women	Mahila Mandal
9.	Farmer	Yuvak Mandal
10.	Farmer	Input Dealer
11.	Farmer	Farmer Group
12.	Block Level Panchayat Raj Institute	Member associated with agriculture
13.	BTT Convener	As Member Secretary

women farmers and weaker sections of the society. Composition given below may suitably modified in composition as per their agro ecological situation.

Composition

Chairman shall be elected out of the above members on rotation basis. BTT Convener also acts as Member Secretary to FAC.

Key Functions of FACs

- Act as an agency for providing farmer feedback mechanism.
- Help set block extension priorities and recommend resource allocation across programme areas.
- Recommend Block Action Plan for approval of ATMA GB.
- Review and provide advice to each implementation unit at block level.
- FAC shall meet once in a month during the season and quarterly in lean season.
- Help in formation of Farmer Interest Groups at block level and below.
- These committees would review and provide advice to each implementation unit at block level.

FARMERS ORGANISATIONS (FOs)

FOs would be encourages at village level and village level groups would, in turn, evolve into Commodity Associations (CAs), Marketing Cooperatives and other types of FOs at the block and district level. At village level Farmer Interest Groups (FIGs) and Farmer Associations (FAs) will be effectively involved in the preparation of block action plans. These organization will coordinate in organizing demonstrations, on-farm and adaptive trials and give feedback to the extension and research. Their representatives would be directly involved in the block level FACs and also at Governing Board of ATMA. The GB of ATMA would select and utilize project funds to support one or more NGOs to assist different types of farmers in becoming organized into different types of FOs within the district.

OPERATIONAL MODALITIES

Funding Mechanism

ATMA will have operational flexibility in use of project funding. They will be expected to adapt plan activities at the district level in consultation with the participating entries as necessary in response to unfolding events. The ATMA Management Committee will be authorized to release project funds onwards to the public/private partners in the agreed activities included in the framework of district extension plan and will maintain separate accounts partner-wise and activity-wise. The accounts (audited by Chartered Accountants) and reimbursement claims will then be routed through the Nodal Cell for onward transmission to the GOI.

Planning and Financial Procedures

The FIAC team would prepare block action plans (BAPs) and budgets that would outline extension and farmer training activities to be undertaken during the coming season. These coordinated plans must address key constraints and opportunities outlined within the SREP if they are to be funded by ATMA. In addition, the Convener, FIAC would be responsible for coordinating these proposed block level extension activities and for submitting these proposals to the FAC for review. After the FAC has approved these proposals, then they would be submitted to the ATMA. The Convener, BTT and the Chairman, FAC from each block would jointly present these extension plans to the ATMA Management Committee prior to the submission to the ATMA GB for approval. In case of programmatic disagreements between the AMC and the FIAC, then these issues would be resolved by the GB.

Once a Block Action Plans (BAPs) have been approved by the GB, then the ATMA Project Director would forward a check to the Convenor, FIAC in each block to cover the budgeted cost of approved extension programs. The FIAC would maintain a bank account and funds would be allocated to each FIAC member in implementing their approved program of extension activities. The Convenor, FIAC and Chairman of the FAC would sign all disbursement checks. The Convenor, BTT would be responsible for maintaining complete financial records, including expenditure receipts, for approved extension activities. Also, the Convenor, BTT would periodically submit detailed financial records to ATMA. The flow of funds to individual blocks would be suspended if financial and performance records are nor submitted to ATMA in accordance with agreed upon procedures.

Operational Procedures

All BTT team members would continue to be employed by their respective line departments, but they would function as a multi-disciplinary technology team or working group that would address the four main programmatic thrusts within the SREP in designing and implementing an integrated extension program. Village Extension Workers (like KPS/AEO/Livestock Assistant/Fishery Assistant, etc.) would have prime responsibility for day-to-day program implementation; with FIAC team members assisting with demonstration plots installation, teaching farmer-training courses, and conducting farm field days, other group activities, etc. The goal of this proposed new arrangement is to create an integrated or *single window* extension system.

To the extent possible, developmental activities financed under central and state government schemes would be utilized to demonstrate and support extension and technology transfer activities within the district and block. In the long-term, the goal would be for more of these central, state and district funds to be directly transferred to the ATMA in support of SREP and BAPs implementation. In the short run, however the FIAC, in consultation with the FAC, would determine where these development activities (especially for agriculture and horticulture) could be most effectively utilized in support of on-going block level extension programs.

Policy Parameters Governing the Cafeteria

In order to ensure that key reforms under the scheme are adequately addresses, notes given at the bottom of the cafeteria specify the following policy parameters within which the cafeteria is to be used:

- *Multi agency extension strategies:* In order to ensure promotion of multi agency extension strategies, minimum 10% of allocation on recurring activities at district level is to be used through non-governmental sector, viz. NGOs, Farmers Organizations (FOs), Panchayati Raj Institutions (PRIs), Para-Extension Workers, Agripreneurs, Input suppliers, Corporate Sector, etc.
- *Farming system approach:* The activities specified in the cafeteria are broad enough to ensure extension delivery consistent with farming system approach and extension needs emerging through SREPs.
- *Farmer centric extension services:* The cafeteria provides for group based extension as it has necessary allocation for activities related to organizing and supporting farmers groups. In order to supplement these efforts, a provision for rewards and incentives to the best-organized farmer groups has also been provided in the cafeteria.
- *Convergence:* The SREP will also be a mechanism for ensuring convergence of all activities for extension. At present, resources for extension activities are being provided under different schemes of Center/State Governments. However, under the scheme, it is being mandated that the work plan to be submitted by the State Governments for funding under the scheme shall explicitly specify the activities to be supported from the resources of other schemes as well as from the proposed scheme.
- *Mainstreaming gender concern:* The gender concerns are being mainstreamed by specifying in the cafeteria that minimum 30% of resources on programmes and activities are utilized for women farmers. Similarly, 30% of resources meant for extension workers are proposed to be spent for women functionaries.
- *Sustainability of extension services:* With a view to ensure sustainability of extension services, it is being mandated that minimum 10% contribution should be realized from beneficiaries with respect to beneficiary oriented activities.

Paradigm Shift of Agricultural Extension

Community development approach to extension: Public sector extension has undergone several changes since the early 1950s. Beginning with the Community Development Programme in 1952 through the National Extension Service in 1953, the focus of extension was on human and community development, but there has been a steady progression towards technology transfer, with the Intensive Agriculture District Programme started in 1961-62, followed by the Intensive Agriculture Area Programme in 1964-65, the High Yielding Varieties Programme 1966-67, the Farmers Training and Education Programme 1966-67 and the Small and Marginal Farmers Development Programme in 1969-70.

Transfer of technology approach through T & V: The most significant development was the introduction of the Training and Visit (T & V) extension management system, starting in the mid-1970s. T & V extension was well suited to the rapid dissemination of broad-based crop management practices for the high-yielding wheat and rice varieties that were released since the mid-1960s. The T & V system profoundly influenced extension practices and registered impressive gains in irrigated areas, because of the similarity between the agro-ecological conditions where technologies were generated and where they were ultimately used, and the favourable socio-economic situations and developmental infrastructure for their wider uptake. Indeed, the T & V system played an important role in ushering in the Green Revolution.

Deficiencies/limitations of extension in T & V system
• The extension personnel received less attention
• A whole farm approach was absent
• Lack of whole family approach
• The extension work was based on individual contact farmer rather than group of farmers.

Post-green revolution period: The T & V system well suited to the rapid dissemination of pre-set agronomic practices for the high-yielding wheat and rice varieties, failed to respond to the more location-specific, risk-prone agriculture of the unirrigated tracts. Similarly, extending the system to programmes for natural resource management, sustainable agricultural practices such as integrated pest management, integrated nutrient management and to diversified agriculture such as high value horticulture, livestock activities and fisheries did not meet with success. Nor could the T & V system adapt to the more holistic Farming Systems Approach towards which the new thrust of both research and extension had begun to focus.

Towards a farming systems approach: The extension approaches of the 1950s and 1960s centered around 'farmers' ignorance' as an explanation of non-adoption of agricultural technology and therefore the extension policies remained confined to "extension education" with the key activities related to "teaching". In the decade of the 1970s and 1980s, farm level constraints were considered to be the explanation of non-adoption. The key activities of extension system were confined to 'input supply' for removal of such farm level constraints. The basic philosophy of these extension approaches centered on "technology transfer". By the early 1990's and the completion of the third National Agricultural Extension Project (NAEP), there was growing recognition that the T & V extension approach needed to be overhauled in meeting the technology needs of farmers during the 21st century. First, it was recognised that extension should begin to broad base its programmes, by utilising a Farming Systems approach. For example, attention should be given to the needs of farmers in rainfed areas, and to diversifying extension programmes into livestock, horticulture, and other high value commodities that would increase farm incomes. Secondly, to support and strengthen the Farming Systems approach, issues of financial sustainability, farmer participation in programme planning, and research-extension linkages, marketing and value addition would have to be concurrently addressed. Present day agriculture is defined by key concepts of stability, sustainability, diversification and commercialization. There is need for reorientation of the philosophy of extension from technology transfer mode to technology application.

Extension Policy Framework in the Context of Changing Scenario

In response to the changes, extension organizations are shifting their principal focus from agricultural productivity alone towards sustainable development, where participatory process, action learning, i.e. the human dimension of agricultural and rural resource management are given importance. The growing consensus to create a demand driven technology system, with direct involvement of farmers in identifying problems, establishing priorities, and carrying out on farm research and extension activities, need a farmer friendly institutions or organization to strike a balance between institution supply system and farmers initiated demand driven extension system.

The above observations provide the very logic for institutional reforms for a newer version for the agriculture and rural development realized much long. In due course, in an effort to respond to the new paradigm; countries world wide had adopted a variety of institutional reforms. It may be noted that the types of extension reform being initiated during the last two decades revolved around decentralization, participation and linkage which have been adopted in the countries like Australia, Germany, India and USA. What is new, however, is the extent of globalization and economic restructuring in both developed and developing countries.

The extension reforms have been categorized into under market reform and non-market reforms. The market reform encompasses four major reform strategies like (i) revision of public sector extension; (ii) pluralism; (iii) cost recovery; and (iv) total privatization. The non market reform comprises two major strategies like decentralization and subsidiary delegating responsibility to the lower level of hierarchy.

At present three decentralization directions currently dominate development of agriculture and rural extension. One is to decentralize the burden of extension cost by redesigning the fiscal system and other is to structural reform. Another type of decentralization has come to exist which intends programme management through farmer participatory involvement in programme planning and decision-making and ultimately the farmer to take responsibility of the extension programme.

In the context of meeting the holistic needs of increasing agricultural production, yet do so in a sustainable manner, agricultural extension has a crucial role to play. Reforms in the system envisage an extension service more broad-based and holistic in content and scope, thus beyond agricultural technology transfer. Technology generation and its application will have to focus more strongly than before on the themes of optimization by producers of their available resources, sustainability, coping with diversity by adapting technology more specifically to agro-ecological or social circumstances aimed at creation of a policy environment that promotes profitable, productive and sustainable farming.

The policy document paper prepared by the Extension Division, Department of Agriculture and Cooperation, Ministry of Agriculture, Government of India. Reforms in agricultural extension already initiated and proposed to be undertaken on wider scale are discussed under the following sub-heads:

- Policy reforms
- Institutional restructuring
- Management reforms
- Strengthening research–extension linkages
- Capacity building and Skill Upgradation
- Empowerment of Farmers
- Mainstreaming of Women in Agriculture
- Use of media and Information Technology
- Financial Sustainability
- Changing Role of Government.

Farming systems approach: Policy reforms in Agricultural Extension envisage the replacement of the old single-discipline based, commodity-oriented approach of the T & V system by the Farming Systems (FS) approach. The FS approach considers the farm, the farm household and off-farm activities in a holistic way to take care not only of farming but also aspects of nutrition, food security, sustainability, risk minimisation, income and employment generation which make up the multiple objectives of farm households. The FS approach emphasises that research and extension agendas should be determined by explicitly defined farmers' needs through an understanding of the existing farming systems rather than perceptions by research scientists or extension functionaries.

Multi-agency extension service: For many years agricultural extension was considered the monopoly of the public sector. However, with the wide range of demands for agricultural technology in the changing scenario there is growing recognition that public extension by itself cannot meet the specific needs of various regions and different classes of farmers. The new extension regime recognizes the role of a multi-agency dispensation comprising different strengths. Policy environment will promote private extension to operate in roles that complement, supplement, work in partnership and even substitute for public extension. The three arms of the agricultural extension network are:

Public extension services
- State government line departments operated extension (Departments of Agriculture, Horticulture and Livestock development)
- State agriculture universities based extension (Directorates of Extension, Krishi Vigyan Kendras (KVKs) and Krishi Gyan Kendras (KGKs)
- ICAR extension (Zonal Research Stations/Krishi Vigyan Kendras, Agriculture Technology Information Centres (ATICs), Institute Village Linkage Programme (IVLP), etc.)

Private extension services
- Community Based Organization (Farmers' Organizations, Farmers' Cooperatives, Self-Help Groups, etc.)

- Para Extension Workers (contact farmers, link farmers, *gopals, mitra kisans, mahila mitra kisans, etc.*)
- Agri-clinics and Agri-businesses
- Input Suppliers/Dealers (Pesticides, Seeds, Nutrients, Farm Implements, etc.)
- Corporate Sector (Commercial crops–tobacco, tea, coffee, oilseeds (sunflower), vegetables, seeds, Farm Implements—tractors, threshers, sprinklers, drip irrigation, etc.)

Mass media and information technology
- Print media—Vernacular Press
- Radio, television, private cable channels, etc.
- Electronic connectivity through Computers, NICNET, Internet, V-SAT, etc.
- Farm Information and Advisory Centres (FIACs)
- Private Portals
- Public and private information shops.

Public extension services: Despite the rise of the private sector in the provision of agro-services, such extension will gravitate towards selected regions, crops and sectors where gains are to be appropriated. Pure public goods, economically backward regions, small, marginal farmers and landless labourers will not attract the for-profit private sector. Public extension will therefore continue to play a central role in technology dissemination. For example, public extension should focus its efforts on those *knowledge-based* technologies that are central to farmers' concerns and that will maintain the natural resource base. These are subject matter areas that are not likely to be taken by the private sector.

Promotion of farmer participatory approach: The extension agent is no longer seen as the expert who has all the useful information and technical solutions; the indigenous technical knowledge of farmers and their ingenuity, individually and collectively, are recognized as a major source; and solution to local problems are to be developed in partnership between the extension agent and farmers. Extension workers therefore need new skills of negotiations, conflict resolution and mobilizing and nurturing community organizations.

Promotion of demand-driven and farmer-accountable extension: Under the T & V system the technology dissemination regime was more supply-driven. Research and extension agendas were pre-set based on technologies for high-yielding varieties of wheat and rice. An important reason why research and extension organisations have not focused on farmer problems is due to the lack of an effective feedback system. The vast majority of small and marginal farmers in India, especially women, lack an effective voice in influencing, research and extension priorities. Under the new policy agenda a demand-driven extension system will be created, by providing farmers with access to linkage mechanisms through which they would be provided all relevant information/data to help them articulate their problems and needs with reference to their production and marketing plans. A key factor in improving these feedback systems is to organise farmers into functional groups, such as Self-Help Groups (SHGs), Farmer Interest Groups (FIGs), Commodity Associations (CAs), and/or other types of farmer organisations (FOs). These FOs can provide an effective channel for both the dissemination of technology to large number of small and marginal farmers and feedback to research and extension.

Thrust on marketing extension: Farmers have increasingly begun to perceive marketing rather than production as the major constraint to enhancing farm incomes. With major thrust of extension agencies on production techniques, marketing extension so far has not received the attention it deserves. This assumes greater significance in the light of the new international trading regime under the WTO and the export opportunities being opened up. Marketing extension so far a peripheral issue in the extension scenario will need to be brought centre-stage. Indeed, production will now need to be significantly dictated by market requirements.

Enabling farmers for problem-solving skills: Under the new dispensation there will be a paradigm shift from top-down blanket dissemination of technological packages, towards providing producers with the knowledge and understanding with which to solve their own location specific problems. This means that the existing public organizations should improve their efficiency and effectiveness in research and technology application. This will call for interdisciplinary approach aiming at location-specificity of technical solution.

Encouraging private sector involvement in technology transfer: A significant deterrent to expansion of private sector involvement in technology transfer is the provision of subsidized agro goods and services by public agencies. This often leads to the creation of an uneven playing field and discourages market entry by private providers. Wherever possible such subsidies will be phased out, in order to stimulate emergence of a private input supply network to provide hybrid seeds, artificial insemination services, fertilisers, agro-chemicals, animal feed, machinery and equipment, and other agricultural supplies and services to farmers on a full cost recovery basis. Targeted subsidies may be retained to protect the interest of the poor and vulnerable sections. The new policy agenda envisages withdrawal of the public sector from areas where agro-services can be effectively and competitively provided by the private sector. In such cases the role of the public sector becomes one of facilitator and enabler. This implies moving towards a realistic cost recovery of agro-services by the State.

Public funds for private extension services: Public funds would be made available to NGOs, Farmer Associations, Para-professionals or private foundations for extension work. An environment in which private investment in technology generation and transfer is more attractive will, therefore, have to be created.

Charging for extension services: The emergence of a market for private extension advice or consultancy services will be encouraged. Processors with contracted producers, also commercial suppliers of seed, agro-chemicals, machinery, vaccines, artificial insemination and the like should recover the costs of providing advice to their clients out of profit margins. However, the vulnerable group will need to be protected through targeted subsidies and safety nets.

Institutional Restructuring

It is clear that no one uniform extension system will serve as a panacea to all States. Even within States there will be a combination of various agencies and different institutional arrangements to address needs of differing agro-climatic zones as well as different sections of farmers. A menu *of various models* will be available to the States to select and adapt to their own requirements.

Restructuring public extension: Public Extension will continue to remain central to Technology Dissemination, small and marginal farmers and economically backward regions will need to be serviced by it. This implies that public extension functionaries (including VEWs and SMSs) will have to be placed in new decentralized institutional arrangements which are demand-driven, farmer-accountable, bottom-up and have a Farming Systems Approach (broad-based). With supplementation from the private sector, media and Information Technology the public extension service would be made leaner and professional.

District Level Agriculture Technology Management Agency (ATMA) Model: A key concept is to *decentralize decision-making* to the district level through the creation of the ATMA as a registered society. A second goal is to increase farmer input into program planning and resource allocation, especially at the block level, and to increase accountability to stakeholders. A third major goal is to increase program coordination and integration between departments so that the following program thrusts can be more effectively and efficiently implemented.

Strategic Research and Extension Plans (SREPs) through Participatory Rural Appraisal (PRA): In the process of creating a more bottom-up extension system, PRA procedures would be introduced across all system levels (district, block, mandal, and village), and across each participating line department (DOA, DOH, DAH and Department of Marketing) and research institutions (ZRS and KVKs) within the district. On the basis of PRA, Strategic Research Extension Plans would be prepared for the districts. The district SREP must be *grounded* at the block or mandal level, where extension programs can be fine-tuned to the needs of farmers and more effectively implemented. The SREP would take account of the research, training and extension requirements for production as well as marketing activities. The rural periodic markets and wholesale assembling markets where farmers visit regularly would be used as important locations for dissemination of market and production technologies.

Block/mandal level technology centre for single window extension system: Concept of a Block (or Mandal) Technology Centre (BTC) has emerged wherein a multidisciplinary technology team (comprising block level agriculture, horticulture, soil and water conservation, agricultural marketing and livestock extension officers) would be assigned to organise and implement extension programs within the block. Other line department units and personnel would continue to provide essential service in developmental activities. In effect, the BTC would result in the functional integration of extension activities within the block or mandal and, in effect, become the operational arm of the ATMA. This Centre would become the common meeting point for extension personnel from the line departments to prepare integrated work plans (WPs) and to coordinate their implementation. It would also be the level where farmer input could be more effectively mobilized through a single Farmer Advisory Committee (FAC). The FAC would include 10–12 members representing all major stakeholders within the block. The FAC would help set block extension priorities and recommend resource allocation across program areas. The block technology team would be responsible for operationalizing the SREP in each block and moving toward a *single window extension system*.

Upgrading and restructuring the extension staff–field extension functionaries as farm advisors: The DOA's extension field staff would be restructured and upgraded to create a professional cadre of Farm Advisors. In the process, the village extension worker (VEW) cadre would be incrementally phased out through reassignment and normal attrition. Eventually, these farm advisors would be incharge of all extension activities within the block and they would all be required to meet a minimum educational requirement for service entry. By the end of this project, this new cadre of extension professionals should be able to identify and provide useful advice for most farmer problems (i.e. become more demand-driven). First, they should be able to carry out a systematic need assessment programme to prioritize farmer problems. Then, by utilizing the strengthened cadre of research and extension specialists (SMSs) within the district, they would be expected to organize and deliver a broader range of extension and farmer training programmes.

Group approach to extension: The *contact farmer* approach to extension popularised by the T & V is to be replaced by the group approach. Formation and mobilisation of Farmer Interest Groups (FIGs), Farmers Cooperatives and Self-Help Groups will be encouraged with the support of NGOs. Group extension will help to replace the top down approach with bottom up approach in technology transfer, as FIGs will first generate a demand for information, technology and management techniques. The extension workers would then respond to the group demand. This would lead to a farmer-extension worker participatory process with emphasis on problem-solving rather than disseminating routine messages. The group approach in extension would also be in line with the Self-Help Groups of rural credit delivery, water user associations, and cooperatives.

Strengthening research-extension-farmer linkages: There is a need for close interaction between farmers, extensionists and production systems researchers in diagnosing problems together and working out location specific recommendations emphasizing participation education rather than prescription and joint actions in the field. Accepted to be more knowledge intensive, these new recommendations will require greater skills both to develop and to apply. There will be strengthening of research-extension-farmer linkages not only at the state and SAU levels but also at the district level; not only between the DOA and the SAUs, but between DOH, DOS and DAH for horticulture, fodder, agro-forestry and silvi-pasture as well as on-farm land and water management in a farming systems approach with due coverage of agricultural marketing concerns. The research-extension interface at all levels from district to national level will be supported.

Promotion of multi-agency extension services: Widening the range of extension delivery agencies for the resource poor farmers and those residing in the hilly, tribal and remote areas, the public system will have to remain as the chief extension mechanism with NGOs possibly being able to play a significant role.

ICAR role in extension—Transfer of KVKs to DAC. The primary mandate of ICAR is research. Its extension programmes should be limited to reinforcing the research activities to make them more demand-driven and farmer centric. Krishi Vigyan Kendras would continue to operate in pro-active mode, retaining their allegiance with ICAR, for project implementation activities. Apart from focusing on production related issues, ICAR research would adequately address different components of marketing and make available need based packages in consonance with the changed/changing agricultural marketing scenario. Links with KVKs will be strengthened at the district level through institutions such as ATMAs.

Management Reforms in Agricultural Extension

Central government support to state governments for extension services on their undertaking policy and institutional reforms. After the close of the World Bank supported NAEP, central support to the state extension services dried up, leaving them with the operation and maintenance of personnel and infrastructure created under T & V. States have barely been able to pay the salaries of extension personnel. Less than 10% of the budget is available for operational expenses, which has practically immobilized the service with scarcely any technology dissemination in the field. It is proposed to support the state extension services provided policy reforms and institutional restructuring is undertaken with demonstrated ability to be demand-driven, farmer-accountable, sustainable and farming systems based with broad-based integrated delivery. While funding for salaries of public functionaries will continue to be the responsibility of the State Governments, funds for technology dissemination and application (operation & management) would be shared between States and Central Government.

Central government funds to be pooled at ATMA or ATMA like registered agencies at district level. Funds from the central government together with state share for all technology transfer and extension activities would be pooled at these district level agencies and released for various activities according to the Strategic Research and Extension Plan prepared for the district.

Central government assistance to state agriculture universities for expanded role in field extension. On the pattern of the successful scientist-farmer-extensionist model developed by the Punjab Agriculture University, the Directorates of Extension of SAUs would be supported to play a larger role in providing extension services in their service-areas.

Central government assistance to KVKs. Under the present arrangement the ownership and mainstreaming of KVKs with the state extension mechanisms has been weak. KVKs, set up as centres for location specific, adaptive research, if effectively organized to achieve their primary objective of refinement and validation of local technologies could play a strategic role in linking the research and extension systems particularly in the area of farming systems based technologies. It is likely that State Governments will be more willing to own and mainstream KVKs once their relevance as district level technology refinement institutions integrated with the extension machinery is demonstrated rather than as just another vocational training organization, which they are largely perceived as at present and of which there are several others at the district level.

Promotion of community-based private extension services. Group approach is the cornerstone of the restructured extension mechanism. A major component of extension services will be the mobilization of the community into farmers groups—FIGs, FOs, and SHGs. Farmers' Organizations will be linked with Panchayats through existing statutory institutional arrangements such as the Land Management Committees, Development Committees, etc. Representatives of FOs would be members of decision-making bodies such as ATMAs, Block level Farmer Advisory Committees, watershed associations. Ultimate aim is for FOs to internalize extension services for its members and provide backward (inputs, credit, technology) and forward linkages (post-harvest facilities, markets, value addition) in a vertically integrated arrangement.

Promotion of NGOs based private extension services. Strength of NGOs is in their ability to mobilize communities into Farmers Organizations/Farmer Interest Groups/ Watershed Associations/Market Associations. As such NGOs complement the public extension effort in several centrally sponsored programmes. Also extension services are contracted out and out-sourced to NGOs at the Block level in some states. A systematic training, capacity building and technical backstopping mechanism, supported through public funds is to be developed for NGOs involved in providing extension services.

Promotion of para-professional based private extension. Para-extension workers normally supplement public extension in a relatively cost-effective manner and overcome constraints of absentee public extension functionaries (Gopals for AI services, Mitra Kisan for agri-services such as soil testing, etc.). Under the new policy agenda para-extension workers at grassroot level will be supported through publicly funded training and capacity building and payment of honorarium in the early years. The honorarium will be routed through the Farmer Organizations/Farmer Groups serviced by the para-extension workers to ensure accountability to the client group. Once the para-worker is able to demonstrate his/her usefulness to the client group the honorarium provided through public funds will be phased out and the client group would take on the onus of paying for the services of the para extension worker. There will be an element of partial/full cost recovery of services provided by para-workers who must ultimately become economically viable units except in the case of vulnerable clients where the State may continue the targeted subsidy.

Panchayati Raj institutions and extension services. After the 73rd Amendment most states are conducting regular elections to the Panchayats. Some states have also delegated suitable administrative and financial powers to the three tier Panchayati Raj institutions. In these states the extension personnel are placed under the administrative control of the panchayats, whereas for technical guidance they remain under the control of their respective technical line departments. Since the panchayat systems are evolving in different states and are currently in a state of flux, the ATMA model at the district, the BTCs and FACs at the Block and the FOs at the village level may be organized as conceived, and suitable linkages be forged with the Panchayati Raj Institutions.

Competitive agriculture extension grant fund. Similar to the Competitive Agriculture Research Grant Fund set up in ICAR and several state governments, wherein both public and private sector research institutions compete for funds to address specific research problems, it is proposed to set up a Competitive Agriculture Extension Grant Fund. Resources under this fund could be accessed through a competitive bidding process. Contracting out extension services to private sector, community-based organizations or NGOs in selected geographical areas (e.g. a village, cluster of villages, Block) would be done through a transparent, laid out procedure under this Fund. This would also imply a strict monitoring and evaluation process.

Linkage of performance with funding for public sector. In a manner similar to the private extension agencies who must compete with one another to access funds and whose subsequent eligibility to compete for funds will depend upon their performance as indicated by an independent impact evaluation, it is proposed that on a pilot basis Public extension agencies also be made to compete with private extension agencies for operational funds under Competitive Agriculture Extension Grant Fund (CAEGF).

Contracting out extension support services. Wherever possible extension services in whole or in part could be contracted out for greater cost effectiveness. This applies, in addition, to administrative services such as security, mobility, computer and secretarial services, participatory planning to NGOs (being done in watershed management), staff training to a University/Institute, monitoring to a Farmer Organizations/IIM/Other Institutions.

Improving Research-Extension Linkages

Promotion of direct interface between farmers and scientists. The direct interface between scientists and farmers is the most ideal and should be undertaken wherever possible. It is an oft-repeated refrain that farmers learn best from scientists or other successful farmers. Moreover, transmission losses are minimized in the direct interface. However, there are relatively high costs attached to this direct mode of technology transfer and the outreach of scientists is limited. Punjab Agriculture University has achieved significant success through this system. It must be noted however, that Punjab is a small state geographically and what is applicable to Punjab may not be possible in large states such as Uttar Pradesh, Madhya Pradesh and Maharashtra.

Activating existing interface mechanisms. Regional Committees of the ICAR, Zonal interfaces initiated by DAC, national level pre-kharif and pre-rabi DAC-ICAR interface, state level bi-annual meetings between line departments and SAUs are all formally instituted mechanisms for improving research-extension linkages. Several of these mechanisms have fallen into disuse or are conducted in a perfunctory manner. As a result the desired results are not being achieved. These will be activated.

Research priority setting based on SREP. Micro-level extension strategies reflected in the Strategic Research and Extension Plans (SREPs) based on PRA and developed jointly by the district technology teams including the marketing department officials and scientists of the KVKs/ZRS or SAUs should be formally feedback into the research systems through a research priority setting mechanism in the ICAR.

Capacity Building of Extension Functionaries

Formulation of HRD policy by states: Central Government support for HRD in Agricultural Extension would be available to the states only after the formulation and adoption of a HRD Policy and Action Plan through a systematic skill-gap analysis. It would also build in an effective system of rewards and incentives for public extension functionaries.

Formulation of training plan for extension functionaries: A long-term training plan should be developed by each state based on a thorough skill gap analysis. A massive campaign will need to be launched for skill upgradation and capacity building of extension functionaries using resources of all training institutes. The training be divided into *Foundation Courses* comprising skill upgradation in (i) need assessment techniques including the role of participatory rural appraisal, (ii) group formation, (iii) development of entrepreneurial skills for agri-business, (iv) agri-business management, (v) WTO and its implications, (vi) marketing of agricultural produce, (vii) post harvest management, (viii) conflict resolution and negotiations between different interest groups, (ix) management of common property resources, (x) use of different type of media, (xi) communication, (xii) project preparation, (xiii) data collection, analysis and documentation.

One time catch-up grant for training infrastructure: One-shot upgradation of physical infrastructure of training institutes/centres be considered to revive the training institutes to an acceptable level. Funding for this purpose to be made jointly by the central and State governments.

Upgrading state level extension management training institutions: Central government would support State Governments to upgrade and restructure their apex state level training institutions to respond to the changing requirements of extension, training and communication management; these upgraded state level apex institutions could have institutional links with MANAGE/NIAM and function as the state arms of the National level Institute. Use of mass media communication techniques will be developed to communicate messages about available technology. Appropriate curricula will be developed for training of field staff, with major focus on marketing related issues.

Strengthening role of MANAGE: The National Institute of Agricultural Extension Management (MANAGE) will be strengthened to enable it to assist the States in developing their HRD capacities.

Developing professionalism in cost effective manner: Training institutes/centres may focus on developing core competency; other services may be out-sourced or contracted. Feedback from participants must be used to evaluate performance of faculty.

Training institutes and SAUs to train private extension functionaries: Facilities of public training institutions and SAUs would be available to NGOs and private extension agents.

Networking among all state level institutes: All national and state level training institutes will be networked to State headquarters, SAUs and MANAGE. The network will also include private institutions with expertise in different fields.

Empowerment of Farmers

Involving farmers in setting extension agenda: Farmers' representation as major stakeholders will be ensured in all decision-making bodies of public and private extension services. Farmer will be involved in the planning and implementation of extension programmes through formal institutional mechanisms such as ATMAs, FACS, etc.

Implementation of programmes through Farmers' User Groups. By ensuring that all programmes in the field are planned and implemented through farmer user groups, such as Watershed Associations, fruit/vegetable growers cooperatives/societies, Agricultural Produce Marketing Societies/Cooperatives, etc. farmers would be able to influence both administrative and financial decisions.

Contracting arrangements between governments, extension services and farmers, whereby the farmers could play the role of beneficiaries, provider or co-financier of extension services.

Acquisition of skills by farmers: Training and acquisition of skills by farmers is a central part of the technology transfer system because of the new practices involved in production. Greater focus will be provided for (i) assessing farmers' needs and skills; (ii) distinguishing different dimensions of training such as awareness, knowledge, skills and reinforcement, and using appropriate channels and methods for each; (iii) different kinds of technologies and advice required by different categories of male and female farmers, the transfer mechanism (e.g. face-to-face, mass media, different types of groups) they prefer during different phases of awareness, trial and adoption of new skills and technologies, (iv) use of information technology for improving the quality and accelerating the transfer and exchange of information; (v) organising training programmes on system based and sustainable technologies such as Integrated Pest Management (IPM) and Integrated Plant Nutrient Management (IPNM); (vi) organizing training and taking initiatives for capacity building of farmers towards agricultural marketing. Capacity building, skill upgradation/training of farmers would be largely conducted through farmers' field schools with an active participation of scientists and extension personnel.

Mainstreaming Women in Agriculture

Mainstreaming women in agriculture: Gender concerns need to be mainstreamed in the agricultural extension process. Public extension systems, which must disseminate new technology and information, are still largely male dominated. Hence the necessity to target women is to ensure that they receive information relevant to their work, particularly, with reference to crops and livestock.

Improving access to extension and training: Women farmers usually have been neglected in extension efforts. Gender inequality had so far not been challenged by the agricultural extension system in the past. However, with the changing scenario, the need for innovating changes in extension approaches has assumed centre-stage. Under these innovations efforts will need to be made both by the central and state governments to improve extension services to reach farm women through (i) extension policy reorientation that explicitly recognizes farm women as agricultural extension clientele; (ii) training of men and women extension staff on women's role in agriculture and rural development and how agricultural extension work could be organized and conducted to meet women's needs in agriculture and rural development activities; (iii) training of women on decision-making in the context of farm and home management, (iv) training of women farmers on agricultural marketing, particularly with respect to post-harvest processing, on farm value addition and market requirements/demand.

Redesign of extension services to reach women farmers: Extension services are being redesigned to focus on women through (i) appropriate training/sensitization of extension personnel towards the role and contribution of women in the total agriculture system, (ii) increasing the proportion of trained female extension workers to gradually ensure that at least one-third of all extension workers are women, (iii) sensitising male extension workers to the needs, approaches and perspectives of women through appropriate training and orientation programmes thereby dispelling the notion that only women can address extension needs of farm women, (iv) improving communication between women, researchers, marketing agencies and extension workers required for the development of technology suitable for women, (v) developing appropriate extension methodologies that recognize the multi-dimensional role of women and the sociocultural barriers, in which women farmers operate in a rural society, (vi) establishing Head of the Farming family as the target group, for extension services and assuming that the information will automatically trickle down to women farmers.

Expanding the sphere of women extension workers: The number of female agricultural extension workers would be increased through (i) re-examination of all service cadre rules for hidden gender biases, (ii) improvement of female attendance at agricultural institutes and school, (iii) building incentives such as scholarships and stipends for more women to take up undergraduate and postgraduate courses in the agricultural and allied sciences, (iv) redesigning of agricultural training curricula to include women's concerns, (v) ensuring that women are adequately represented in all training programmes whether domestic or overseas, (vi) redesigning of training facilities to make them suitable for large numbers of female students and trainees, (vii) inclusion in the teaching curricula for extension workers, greater analysis and extension methods that take into account women's time, mobility and cultural situation; and (viii) exploring the specific role of farm women in the marketing of agricultural produce.

Use of Information Technology

Information technology revolution is unfolding, and has very high visibility. However, its benefits have remained confined primarily to the urban areas. Rural communities have not been able to gain to the same extent from IT. As a means of agricultural technology transfer to farmers, information technology, has had a limited impact. Even the vast potential of the broadcasting network has been tapped only minimally for extension.

Increasing use of information technologies: Harnessing information technology for agricultural extension will receive high priority in the new policy agenda. Extensive use of modern information technology will be promoted for communication between researchers, extension workers and their farmer clients to transfer technologies and information more cost effectively. Information technology should be made available, particularly, to those with specific inquiries to guide them in adopting the more knowledge intensive forms of agriculture, which will expand in future.

IT application in agriculture marketing: Agriculture produce marketing requires connectivity between the market and exporters/growers/traders, industry consumers, through wide area network of national and international linkages in order to provide day-to-day information with regard to commodity arrivals and prevailing rates, etc. to provide links for online International market information; to provide export related documentation, to inform about the latest research in agricultural marketing, packaging, storage related information and to provide connectivity with lead international and national market organizations.

Wider use of electronic mass media for agricultural extension: Radio and TV have vastly increased their reach, as also reception facilities. "Local" radio and new FM transmitters open up possibilities of area-specific broadcasts. In communicating with an audience with low literacy skills, an audio-visual medium like TV has advantages. Today Doordarshan covers the entire population. Much wider and creative use of the mass media—All India Radio, private FM, Doordarshan, private cable network will be promoted for more rapid and effective dissemination of general information and advice to farming communities.

Farmer participation in IT programmes: In developing any system of IT for agriculture technology transfer; the farmer is to be kept centre-stage. She/he is not to be treated as a passive recipient but rather as a player, a generator and user of knowledge. The upgradation of his/her skills and knowledge is therefore a crucial part of the process. The farmer will be an effective participant in the process.

Support to states for information technology: Increased use of information technology at State/district and block levels would be promoted. This would include electronic access, through NICNET, to technical and administrative information. Central Government will support States in the use of electronic linkages and computerization so that marketing, research, extension and farming communities are linked to each other, and into local, national and global networks.

Private information shops/kiosks: The ultimate aim is to promote private Information shops/kiosks franchised out to private sector especially unemployed rural educated youth, in the manner of PCOs/STD shops. Private sector will be encouraged to establish information shops at Block/Mandal/Village level. A major programme for development of software will need to be mounted so that Information Shops could have access to suitable material. Electronic connectivity and access to e-mail would put the franchisees in contact with district KVKs, Line Departments, markets and other sources of information. Such information could be dispensed to farmers, farmers groups on payment.

Capacity building for use of IT: Application of IT is constrained by lack of or inadequacy of complementary inputs (equipment, power, etc.), appropriate organisational and institutional structures, information management and skills development. A major training programme for developing capacity for IT usage will be promoted. Training Institutes will run suitable courses for the purpose.

Financial Sustainability and Resource Mobilization

Publicly funded extension will continue to play a predominant role in technology dissemination firstly because the large numbers of small disadvantaged farmers may not have access to or be able to afford any other kind, and secondly, because much of the new technology will not be commercially marketable for instance watershed management, land capability assessment and land use planning, breaking of yield ceilings sustainable management of natural resources and socioeconomic research. But pressures on government expenditure mean that public funds will have to be more carefully targeted and more efficiently used.

Cost-cutting mechanisms for extension services: Cost effectiveness may be improved by relying on fewer but better qualified (graduate or postgraduate) field advisers who interact directly with researchers for subject matter advice and then multiply their impact in the field by working with farmer groups rather than individual contact farmers. Cost cutting mechanisms, including the exploitation of mass media, encouragement of NGO and private sector involvement in extension, or needs-based coverage.

Efficient use of available resources: Optimum fund utilization will be achieved with better matching the farmers needs with extension delivery, a stronger focus on the economics of farming, and the use of participatory methods to assess needs, create commitment to action, and monitor impact.

Privatization of agro-services: An environment in which private investment in technology generation and transfer is more attractive will be created. Product diversification both horizontal and vertical shall be promoted to not only improve profitability sustainability and more efficient use of production resources but also to encourage greater involvement of the private sector. Where opportunities exist to contract out publicly funded services, or to transfer costs to the corporate sector or to users themselves, these opportunities should be exploited for instance for diversification into higher value or export crops, or to develop new commercial inputs or machines. Privatization of selected "private goods" and agro services wherever a competitive market exists, such as AI services, soil testing, fertilizer advice, farm improvement plans or breeding plans would be undertaken.

Towards a realistic cost recovery of agro-services: Wherever farmers have the capacity to pay for public services, which are in the nature of private goods, realistic cost of such services should be recovered. However, provision is made for targeted subsidies to protect the vulnerable class of users.

Co-financing of public extension: Co-financing of public extension services by farmers and farmers' associations to reduce pressure on public finances and to improve the accountability and responsiveness of extension to farmers.

Initiating new financial systems: Modification in rules and regulations and innovations in financial institutions will also be required to allow for arrangements such as "revolving funds" for government farms, nurseries, etc. While budgetary support to these units may continue to meet pay and allowances of government staff, the funds for recurring expenditure and operational costs could be generated by these units from the commercial activities undertaken by them. All efforts would be made to develop credit linkages of farmers and farmers groups with credit institutions.

Changing Role of Government

Role of state in effective regulation and enforcement: As a multi-agency extension regime proliferates, the responsibility of the State for effective enforcement of legislation, which ensures quality control of inputs such as seed, pesticides, fertilizers, etc. will increase. State's role as arbitrator of conflicts between various private sector extension agents will also increase and systems to address grievances will need to be developed. This role will increase as the number of private extension agencies grows. Guidelines for private agencies would be required. However, in the emerging pluralistic scenario the role of public extension would need to be redefined from one of solely a provider of services to become increasingly an appropriate mix of provider, coordination, facilitator and regulator. The large section of small and marginal farmers and landless labourers as well as remote and backward regions would continue to need the services of the public extension functionaries, as they are not likely to be serviced by a competitive private sector in the near future. Public Extension's role would increase in arbitration of conflicts, assuring accountability of all service providers to the farmers and ensure transparency through provision of information. The overall environment of private provision of extension services deserves to be encouraged through policy reforms and institutional changes so that rural people's needs serviced more efficiently.

Creating an enabling environment: Generally, this implies appropriate legislation, rules and regulations, and application of the rule of law. In particular, it implies that private contracts and property are protected and a judiciary exists to enforce contracts without partiality and undue delay. Where many individual smallholders are concerned a legal course of action may, however, not be practical or politically expedient for handling conflicts and disputes. Government can instead support institutions like an independent arbiter, an ombudsman, or a small farmer reference service that would certify bona fide borrowers or agricultural producers. Governments can also set minimum standards and norms for commodities such as food, pesticides, and packaging materials when it is in the interest of public health. To protect the weaker of the contracting parties, governments can propose minimum standard contract clauses and guidelines for small farmer/agri-business transactions. It is essential that such proposals be seen as recommendations, not prescriptions.

Enhancing competition: The enhancement of competition is another government contribution to improving the institutional environment. It involves all measures to ensure open, fair and transparent competition and to facilitate entry of newcomers. It may include breaking up of monopolies and cartels, ensuring minimum professional standards of business conduct, and resisting demands for non-technical obstacles to official licensing by rent-seeking lobbies. A lack of financial means is frequently the reason that prevents newcomers from establishing new businesses. Governments can assist young entrepreneurs to access credit and venture capital by providing technical assistance to prepare business plans, conduct market surveys and hire help to resolve special engineering or legal problems and through this to improve confidence of funding sources in new ventures. Part of creating an enabling environment would also be to address the downside of privatization and liberalization.

Strengthening farmers' associations: Government services can help identify existing associations or cooperatives of farmers and support them to develop their organization.

The aim must be to assist the groups to define their objectives, such as savings mobilization or specific post-harvest operations, to ensure group coherence and continuity, and to assist them with setting up group structures and organization. Over time such groups can establish a track record of organizational maturity that will allow, possibly after joining with other groups for economies of scale, to engage in their own business activities and to gain access to formal credit. Government extension services and NGO staff need to receive suitable training to act as group facilitators. Support to farmers' organizations is perhaps the main single input that governments can provide for the promotion of farmer integration with agri-business.

Strengthening physical infrastructure: The government's role would increasingly be in the area of physical infrastructure provision, including communications and utilities, farm-to-market roads, and rural and urban markets. Promotion of private sector would be through making available sites with road and electricity connections to attract enterprises that may set up marketing or processing facilities as part of an industrial estate. Rural or farm-to-market roads also facilitate linkages between farmers and private service providers. Similarly development of wholesale market yards is also supported by governments at given stages of marketing development. At more advanced stages such facilities tend to lose their functions, as alternative forms of marketing develop for a variety of agricultural products that rely on direct producer/ agri-business/consumer linkages and by-pass traditional markets.

Improving information: Another way of leveling the playing field for private sector is the improvement of information. Information gathering and analysis is costly. Compared to commercial business, farmers are at a disadvantage on knowledge about prices, volumes, qualities, alternative marketing channels and other feature governing market transactions. Government can improve the communications flow and the quality of information to farmers through training workshops and publications and by this improve transparency and facilitate transactions. Government can also sponsor market-matching exercises, that is, sponsor meetings and workshops involving farmers and agri-business enterprises to improve mutual understanding of constraints and requirements, and promote concrete business deals.

Innovation System vis-à-vis Farm Sector

INNOVATION

As mentioned in Business Dictionary.com, "Innovation is the process of translating an idea or invention into a good or service that creates value or for which customers will pay".

"Innovation refers to new concepts or products that derive from Individual's idea or from scientific research". Innovation, on the other hand is the commercialization of the invention itself as mentioned in Merriam-Webster Dictionary.

Innovation is the process by which organizations " master and implement the design and production of goods and services that are new to them, irrespective of whether they are new to their competitors, their country, or the World" (Mytelka, 2000).

In its broadest sense innovation covers the activities and processes associated with the generation, production, distribution, adaptation and use of new technical, institutional and organizational or managerial knowledge. It does not mean new technology alone, but also the institutional and organizational innovations, that emerge as new ways of developing, diffusing and using technology (Anandajayasekeram et al., 2008).

The capacity for innovation occurs in one or more of four trajectories: Product innovation, Process innovation, Management or Organizational innovation and Service delivery innovation. It is also found that the two factors of importance in innovation are Knowledge and Networking, i.e. value of knowledge increases with its use, and exchange can only be realized in a cooperative environment (Anandajayasekeram et al., 2008).

The concept of innovation has evolved over the last few decades. Historically, innovation was defined as a new technology or knowledge developed by scientists, transferred by extension personnel and adopted by farmers. Governments established extension organizations in different countries mainly to communicate or transmit new technologies developed by agricultural research centres. Subsequently, with the increasing realization that organized research alone is not the main source of new knowledge or technology, the concept was broadened to include knowledge and technology developed by others, mainly farmers or communities. While formally the focus was on technical innovations, it is now recognized that social and institutional innovations are needed to promote agricultural development. Thus innovation began to be defined in terms of technical, social and institutional innovations.

Changing approaches for supporting Agricultural innovation

As the context of agricultural development has evolved, ideas of what constitutes "research capacity" have evolved, along with approaches for investing in the capacity to innovate:

- In the 1980s, the National Agricultural Research System (NARS) concept focused development efforts on strengthening research supply by providing infrastructure, capacity, management and policy support at the national level.
- In the 1990s, the "Agricultural Knowledge and Information System" (AKIS) concept recognized that research was not the only means of generating or gaining access to knowledge. The AKIS concept still focused on research supply but gave much more attention to links between research, education and extension and to identifying farmers demand for new technologies.
- More recently, attention has focused on the demand for research and technology and on the development of innovation systems, because strengthened research systems may increase the supply of new knowledge and technology, but they may not necessarily improve the capacity for innovation throughout the agricultural sector (World Bank, 2006).

However, the role of extension was mainly identified as promoting dissemination of technical innovations. Discussions on the AKIS further broadened this concept, and innovation began to be described as the emergent characteristics of the interactions among stakeholders in a natural resource or ecosystem service (Rolling and Wagemakers, 1998).

Innovation began to be recognized as the outcome of the interaction among the diverse actors addressing a particular problem. In this scenario, the role of extension was seen as facilitating the process of reflective action, learning and decision-making (Rivera, 2001).

Innovation System

An innovation system is the set of organizations and individuals involved in the generation, diffusion, adaptation and use of new knowledge of socioeconomic significance and the institutional context that governs the way these interactions and processes take place (Anandajayasekeram et al., 2008).

The innovation systems concept is attractive not only because it offers a holistic explanation of how knowledge is produced, diffused and used but because it emphasizes the actors and processes that have become increasingly important in agricultural development. To recapitulate some of the points made earlier, agricultural development plans are no longer concerned almost exclusively with staple food production. These plans now give far more attention to diversification into new crops, products, and markets and to adding value to serve new markets better (Bhargouti et al., 2004).

Beginning in 2000, the innovation systems concept had been applied to agriculture (Hall et al., 2001; 2008; The World Bank, 2006). At this time, the AIS framework started to gain increasing attention in the international development community, as it provided an overarching framework for knowledge advancement.

There is no assumption that an innovation process starts with research or that knowledge feeds directly or automatically into new practices, processes or products.

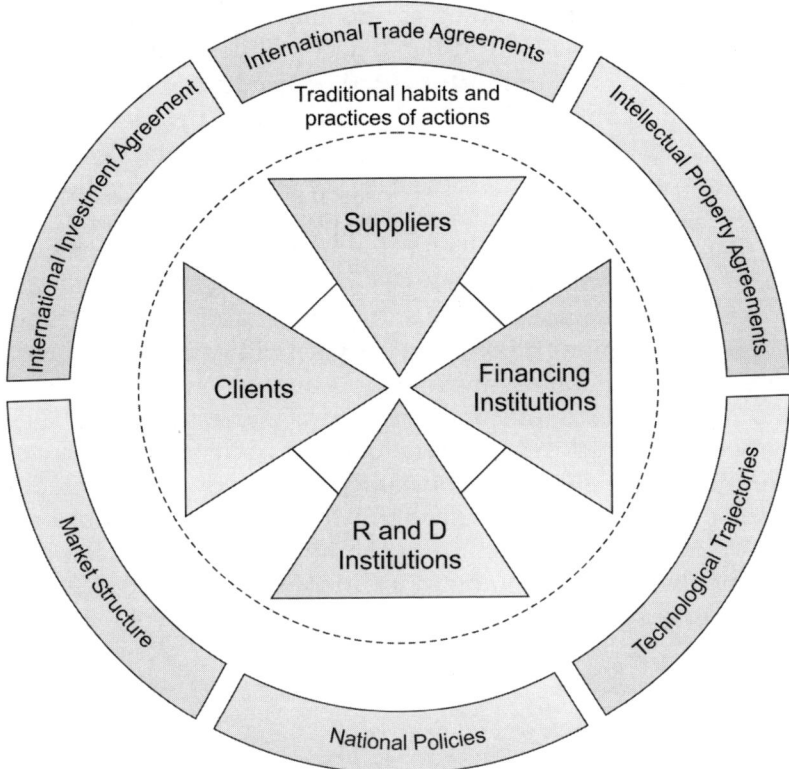

Fig. 9.1: Innovation systems. *Source:* Mytelka (2000)

Instead, the knowledge and information flows at the heart of an innovation system are multidirectional. They open opportunities for developing feedback loops that enhance competence building, learning and adaptation. All too often, the right kinds of actors are absent, or they do not interact in ways that support the innovation process. The innovation systems concept helps to reveal why these interactions might not be present and what be done to remedy this problem (World Bank, 2006).

Because new markets for agricultural products and services change continuously, agricultural development depends more than ever on a process of continuous, incremental innovation. The scope of innovation includes not only technology and production but organizations (in the sense of attitudes, practices, and new ways of working), management, and marketing changes, therefore requiring new types of knowledge not usually associated with agricultural research and new ways of using this knowledge, ways of producing and using knowledge must also adapt and change. In Fig. 9.1, the innovation systems concept emphasizes adaptive tendencies as a central element of innovation capacity.

Agricultural Innovation System

More recently, approach of World Bank has moved towards the concept of "agricultural innovation systems" (AIS) and focuses on strengthening the broad-spectrum of science and technology activity of organizations, enterprises and individuals that demand

and supply knowledge and technologies and the rules and mechanisms by which these different agents interact (World Bank, 2006).

The AIS framework is founded on a rather broad definition of 'innovation', recognizing that innovation is not a research driven process simply relying on technology transfer. Instead innovation is a process of generating, accessing and putting knowledge into use, and it recognizes the need for communication flow among a wide range of actors involved in generation, access and implementation for innovation to happen. Like that Agriculture Knowledge and Innovation System (AKIS) framework, it stresses the need for institutional linkages and learning. Reflecting major trends of agricultural development since the 1990s, this framework further acknowledges the importance of linkages among a broad set of stakeholders, which, include the private sector and non-governmental organizations. It moves innovation into the centre of attention, and focusing on results and stressing a wider range of stakeholders than the AKIS framework. Structures and relationships of institutions are less static, but collaboration and partnerships are constantly reviewed and adopted according to changing needs. This also makes clear that innovations are not only related to technology developments, but include social and institutional components. From, Fig. 9.2 it is essentially, the AIS promote innovation-driven development and pluralistic networking among agriculturally relevant organizations (Rivera, 2001).

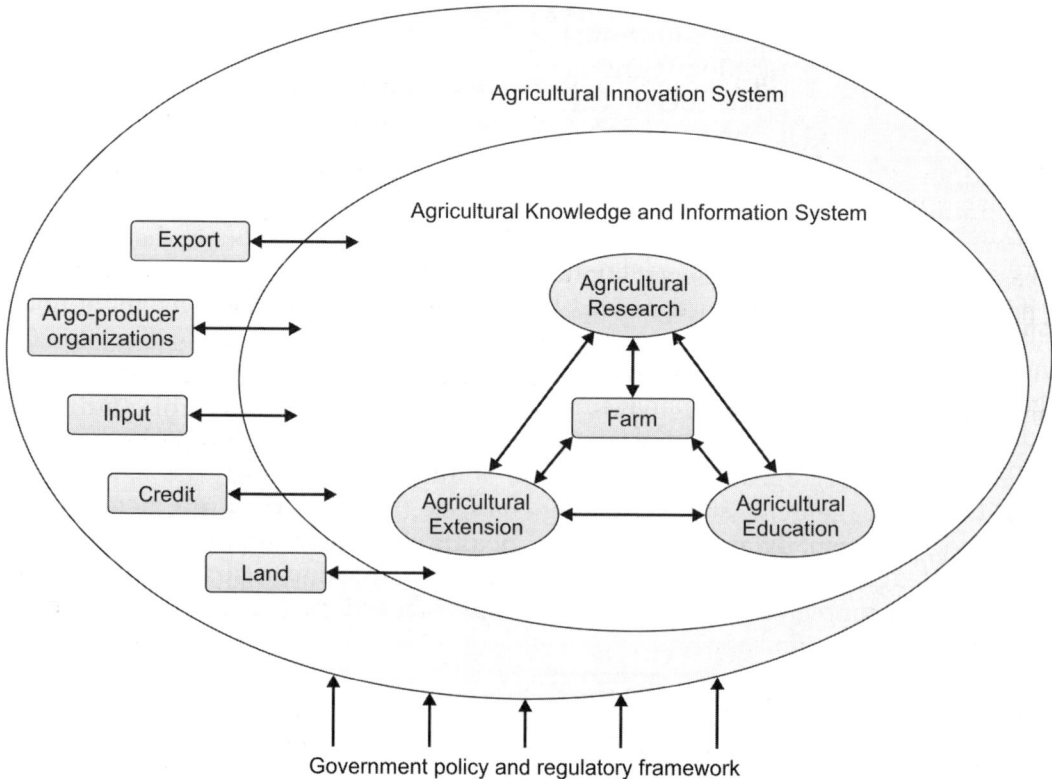

Fig. 9.2: Agricultural innovation system (AIS). *Source:* Rivera, et al. (2005)

Institution

An institution is any persistent structure or mechanism of social order governing the behavior of a set of individuals within a given community. Institutions are identified with a social purpose, transcending individuals and intentions by mediating the rules that govern living behavior (Wikipedia).

According to business dictionary, institution is an establishment, foundation or organization created to pursue a particular type of endeavour, such as banking by a financial institution.

As stated in Cambridge Dictionary, institution is an organization where people do a particular type of scientific, educational or social work, or the building that it uses.

Oxford dictionaries writes, "institution is an organization founded for a religious, educational, professional or social purpose".

According to dictionary.com, institution is an organization, establishment, foundation, society or the like, devoted to the promotion of a particular cause or program, especially one of a public, educational or charitable character.

Generally speaking institutions refer to man-made rules that govern human behavior. The general and common understanding developed by persons and groups, i.e. culture is the basis for the design of frameworks of more specific rules that govern human behavior. Various persons and groups develop different institutions. Families, businesses, government agencies and churches, for instance, all have their own institutions.

According to Adam Smith culture matters when explaining and fostering economic growth. In this theory of 'Moral Sentiments' Smith (1759) analyses what it means for persons to be rational and social. These are the necessary conditions for a well-functioning system of free markets, as analyzed in his Wealth of Nations (Smith, 1776).

In the period that economics increasingly disregarded institutions other social scientist, sociologists in particular put institutions at the centre of their analysis (Durkheim, 1893). Economics and other sectors of society can only function well if there are institutions that hold society together.

Institutional Innovation

"Institutional innovation" or the new ways of doing things includes, new ways of organizing production, input management, marketing or sharing common resources. It could even be the development of new producer company or a new way of providing extension support.

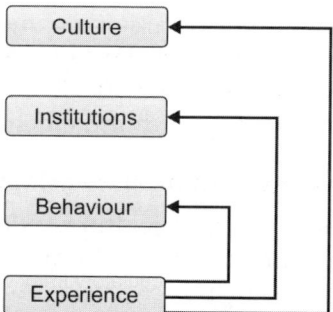

Fig. 9.3: Cultural and institutional change. *Source:* Keizer (2008)

Sulaiman et al. (2006), argued that institutional innovations are going to be equally or more important in dealing with complex challenges facing agriculture and rural development. Technical innovation need not have to be the starting point for extension. Extension has several other things to do even in a situation where relevant technical innovations do not exist. Technical and institutional innovations ideally should go hand in hand. Quite often, institutional innovations are even necessary for generating and promoting technical innovations. Institutional innovations can flourish only in a situation where sufficient flexibility and freedom to experiment exists. However, centralized arrangements for funding, implementation, monitoring and evaluation have been found to effectively stifle generation of locally relevant institutional innovations.

National Agricultural Technology Project (NATP)

The National Agricultural Technology Project (NATP) with the financial support of the World Bank was initiated in the financial year 1995-96. Government of India approved this project in November, 1998 for full-scale implementation. The diverse activity of the project has been planned under three major components.

The development objectives of the national agricultural technology project are to: (1) improve the efficiency of the Indian Council of Agricultural Research (ICAR) organization and management systems; (2) enhance the performance and effectiveness of priority research programs and of scientists in responding to the technological needs of farmers; and (3) develop models that improve the effectiveness and financial sustainability of the technology dissemination system with greater accountability to and participation by, the farming communities. The project components are: (i) Development of the ICAR's organization and management systems; (ii) Support for agro-ecosystems research through sponsored and competitive grants program and (iii) Support for pilot program to test support for innovations in technology dissemination.

Innovation in Technology Dissemination (ITD) Component of National Agricultural Technology Project (NATP)

The ITD component of NATP was started in 1998 with a view to pilot test reforms in agricultural extension with the support of the World Bank. The project was operationalized through Ministry of Agriculture in 28 districts; four districts each of seven states namely—Andhra Pradesh, Jharkhand, Maharashtra, Odisha and Punjab. The project focused on restructuring public extension services and testing new institutional arrangements for technology transfer (MANAGE, 1999).

To operationalized the above reform initiatives under ITD component of NATP, Agricultural Technology Management Agency (ATMA) has been established in each project as an autonomous institution providing flexible working environment involving all the stakeholders in project planning and implementation.

Agricultural Technology Management Agency (ATMA)

It was into the top-down public-sector agricultural extension system that ATMA first emerged in 1998 as a part of the World Bank-funded Innovations in Technology Dissemination (ITD) component of the National Agriculture Technology Project (NATP). It was implemented as a pilot in 28 districts in seven states of India from 1999 to 2003 (Reddy and Swanson, 2006).

Agricultural Technology Management Agency (ATMA) is a registered society of key stakeholders involved in agricultural activities in a district and it is the focal point for integrating research and extension activities and decentralizing the management of the public agricultural technology system. As a society, ATMA can receive and expand project funds, enter into contracts and agreements and maintain revolving accounts that can be used to collect fees and thereby recovering operating costs (MANAGE, 1999).

Swanson et al. (2008) have defined ATMA as a semi-autonomous decentralized participatory and market-driven extension model and represents a shift away from transferring technologies for major crops and towards diversifying output (Asenso-Okyere et al. 2008).

The aims of ATMA are to integrate extension programs across state-level departments, link research and extension activities in a district, and decentralize extension decision-making through participatory planning (Singh and Swanson, 2006). As a registered society, ATMA has more flexible to access funds, provide funds and work in partnership with the private sector than previously (Birner and Anderson, 2007).

Diagrammatic depiction of composition and processes of ATMA given in Fig. 9.4 present the approach and process institutionalized through ITD components as a reform process.

The ATMA is a unique district level institution, which caters to activities in agriculture and allied departments adopting a Farming System Approach. It can receive funds directly (Government of India/States, Membership fees, beneficiaries' contribution, etc). ATMA is be supported by a Governing Board (GB) and a Management Committee (MC). The programmes and procedures concerning district wise R-E activities are determined by ATMA, GB and implemented by its MC (Singh and Swanson, 2006).

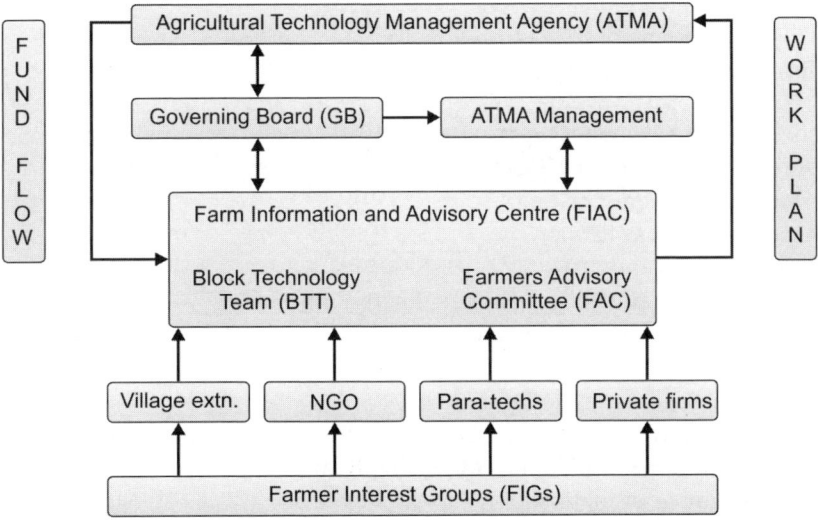

Fig. 9.4: Organizational structure of ATMA. *Source:* Singh and Swanson (2006)

ATMA Governing Board

The Governing Board at district level was the policy making body and provide guidance as well as review the progress and functioning of ATMA.

Table 9.1 depicted that the ATMA GB was a sixteen member committee with 9 official and 7 non-officials. The District Collector as the Chairman heads the GB with district level officials like Agriculture, Credit, and Rural Development, etc. as members along with one member from district level research unit, i.e. KVK. The farmer members representing enterprises like agriculture, horticulture, animal husbandry, fishery, women and SC and ST community have been taken on board.

The key functions of ATMA Governing Board:

As mentioned in the project support document developed by MANAGE, Hyderabad; the objectives of GB were to:

1. Review and approve Strategic Research and Extension Plan (SREP) and annual action plans that are prepared and submitted by the participating units.
2. Receive and review annual reports presented by the participating units, providing feedback and direction to them as needed, for various research and extension activities being carried out within the district.
3. Receive and allocate project funds to carry out priority research, extension and related activities within the district.
4. Foster the organization and development of Farmers Interest Groups (FIGs) and Farmers Organizations (FOs) within the district.
5. Facilitate the greater involvement of private sector and firms and organizations in providing inputs, technical support, agro-processing and marketing services to farmers.

Table 9.1: Composition of ATMA governing board

S. No.	Official and non-official member	Designation/position
1.	District Magistrate/ Collector	Chairman
2.	Chief Executive Officer (CEO)/ Chief Development Officer (CDO)	Vice-Chairman
3.	Project Director ATMA	Member-Secretary-cum Treasurer
4.	Joint Director/Deputy Director (Agri.)	Member
5.	A representative from ZRS/KVK	Member
6.	One Farmer representative	Member
7.	One Horticulture Farmer	Member
8.	One Livestock producer	Member
9.	Representative of Women Farmers Interest Group	Member
10.	One SC/ST farmer representative	Member
11.	A representative of NGO	Member
12.	Lead Bank Officer of the District	Member
13.	A representative of District Industrial Centre	Member
14.	Representative of Agriculture Marketing Board	Member
15.	Representative of input supplying Associations	Member
16.	One Fisheries/ Sericulture representative	Member

6. Encourage agriculture lending institutions to increase the availability of capital to resource poor and marginal farmers, especially SC, ST and women farmers.

7. Encourage each line department, plus the KVK and ZRS, to establish farmer advisory committees to provide feedback and input into their respective R-E Programmes.

8. Enter into contracts and agreements as appropriate to promote and support agricultural development activities within the district.

9. Identify other sources of financial support that would help ensure the financial sustainability of the ATMA and its participating unit.

10. Establish revolving funds/accounts for each participating unit and encourage each unit to make available technical services, such as artificial insemination or soil testing on a cost recovery basis moving towards full cost recovery in a phased manner.

11. Arrange for the periodic audit of ATMA financial accounts.

12. Adopt and amend the rules and by-laws for the ATMA.

ATMA Management Committee

The ATMA Management Committee (AMC) is the apex technical committee formed with 11 members representing line departments, research institutions, one NGO and two farmers. From the below Table 9.2 it is amply evident that the AMC headed by the project Director of concerned ATMA was responsible for planning and execution of the day-to-day activities of ATMA.

Key Functions of ATMA Management Committee (AMC)

The function and tasks to be carried out by the ATMA Management Committee as mentioned in the project support document prepared by MANAGE, Hyderabad were:

1. Carryout periodic Participatory Rural Appraisal (PRA) to identify the problems and constraints faced by different socioeconomic groups and farmers within the district.

Table 9.2: Composition of ATMA Management Committee

S. No	Official and non-official member	Designation/position
1.	Project Director, ATMA	Chairman
2.	District Head of Dept., Agriculture	Member
3.	District Head of Dept., Horticulture	Member
4.	District Head of Dept., Animal Husbandry	Member
5.	District Head of Dept., Fisheries	Member
6.	District Head of Dept., Sericulture	Member
7.	District Head of other appropriate line dept.	Member
8.	Head, Krishi Vigyan Kendra	Member
9.	Head, Zonal Research Station	Member
10.	One representative of NGO in-charge of Farmers' Organization	Member
11.	Two representative of Farmers' Organizations (one year rotation basis)	Member

2. Prepare and integrated, Strategic Research and Extension Plan (SREP) for the district that would specify short and medium term adaptive research as well as technology validation and refinement and extension priorities for the district: these priorities should reflect during the PRA.

3. Prepare annual action plans that would be submitted to the ATMA Governing Board for review, possible modification and approval.

4. Maintain appropriate project accounts for submission to Technology Dissemination Unit (TDU) for audit purposes.

5. Coordinate the execution of these annual action plans through participant line departments, ZRSs, KVKs, NGOs, FIGs/FOs and allied institutions, including private sector firms.

6. Establish coordinating mechanisms at the Block level, such as Farm Information and Advisory Centres (FIACs) that would integrate extension and technology transfer activities at the block and village levels.

7. Provide annual performance reports to the Governing Board outlining the various research extension related targets that were actually carried out.

8. Provide Secretariat to Governing Board and initiate action on policy direction, investment decisions and other guidance received from the Governing Board.

Participation

Participation, in the development context, is a process through which all members of a community or organization are involved in and have influence or decisions related to development activities that will affect them. That implies that development projects will address those community or group needs on which members have chosen to focus, and that all phases of the development process will be characterized by active involvement of community or organization members.

According to Oxford Dictionaries "participation" is the action of taking part in something. The Century Dictionary and Cyclopedia has defined participation as "the act or fact of participating or sharing in common with another or with others; the act or state of receiving or having part of something".

Some people defined participation as any "voluntary or other forms of contribution by rural people to pre-determined programs or project". Activities such as participation in a survey, serving as key informant, or participation in an experiment which is research-managed could be described as participation (Anandajayasekaram, 2008).

The primary aim of participatory approach is to encourage the active participation of rural people in planning, implementing, managing and monitoring extension programs. To achieve this participation, extension organizations will need to formally decentralize or transfer the control of specific program planning and management functions to the local system levels where extension programs are actually implemented (Swanson and Samy, 2004). On the one hand, participation has been defined by one author as 'voluntary or other forms of contributions by rural people to pre-determined programs or project' (Oakley, 1991). On the other hand, a participatory project has been described as one initiated and owned by beneficiaries (Cummings, 1997).

Participation can be considered as a product (end) as well as a process (means). As a product, the act of participation is an objective in itself, and is one of the indicators of success as it refers to the empowerment of individuals and communities in terms of acquiring skills, knowledge and experience, leading to greater self-reliance. However, when viewed as a process, participation refers to the action used to achieve a stated objective, i.e. cooperation and collaboration which helps to ensure sustainability of program/project/development. In the literature, a distinction is made between 'participation' and 'participatory'. The term participatory development has sometimes been defined as involving users and communities in all stages of the development process (Narayan, 1993).

A World Bank (1994) review of development programmes which it supports concluded that people's participation continues to be both complex and difficult. It is more than simply beneficiary participation in economically successful development projects.

Achieving participatory approach entails delegating roles to the local committees (local leaders, farmers' organizations and agricultural credit organization) so that they can contribute to the planning and implementation of agricultural extension programs (Rivera et al., 2006). The participatory approach is a framework for extension staff to participate with farmers in facilitating development planning and activity implementation in a local region. This helps in strengthening farmers' problem-solving abilities from the start. In relation to sustainable agricultural development, the existence of a local management and decentralized administration is a precondition (Okorley et al., 2009).

Decision-Making

Decision-making is the mental process of choosing from a set of alternatives. Every decision-making process produces an outcome that might be an action, a recommendation, or an opinion. Since doing nothing or remaining neutral is usually among the set of options one chooses from, selecting that course is also making a decision (Boundless, 2014).

Decision-making means to select a course of action from two or more alternatives. It is done to achieve a specific objective or to solve a specific problem.

According to dictionaries the meaning of decision-making stand as:
- Choice that you make about something after thinking about it (Merriam Webster Dictionary).
- Thought process of selecting a logical choice from the available options (Business Dictionary).

According to Wikipedia, decision-making can be regarded as the cognitive process resulting in the selection of a belief or a course of action among several alternative possibilities. Every decision-making process produces a final choice that may or may not prompt action. It is the study of identifying and choosing alternatives based on the values and preferences of the decision maker. Decision-making is one of the central activities of management and is a huge part of any process of implementation.

Although cognitive aspects of decision-making are considered important to adolescent risk-taking (Cousins and Rubovits, 1993), risk-related decisions require additional considerations (Furby and Beyth-Marom, 1992). In general, "decision-making" can be defined as process of making choices among possible alternatives. The skill considered important to effective decision-making are based on a normative model of decision-making, which prescribes how decisions should be made. These skill include: (1) identifying the possible options; (2) identifying the possible consequences that follow from each option; (3) evaluating the desirability of each of the consequences; (4) accessing the likelihood of each consequence; and (5) making a choice using a "decision rule" (Furby and Beyth-Marom, 1992).

In the view of FAO, closely related to both strategic and managerial planning is the process of decision-making. Decisions need to be made wisely under varying circumstances with different amount of knowledge about alternatives and consequences. Decisions are concerned with the future and may be made under conditions of certainty, conditions of risk or condition of uncertainty. Decision-making is more than making up mind. It consists of several steps:

Step 1: Identifying and defining the problem

Step 2: Developing various alternatives

Step 3: Evaluating alternatives

Step 4: Selecting an alternatives

Step 5: Implementing the alternative

Step 6: Evaluating both the actual decision and decision-making process.

In extension, the decision-making process is often a group process. Consequently, the manager must apply principles of democratic decision-making since those involved in the decision-making process will feel an interest in the results of the process. In such a case, the manager becomes more of a coach, knowing the mission, objectives, and the process, but involving those players who must help in actually achieving the goal. The effective manager thus perceives himself or herself as the controller of the decision-making process rather than as the maker of the organization's or agency's decision.

As Drucker (1966) has pointed out, "the most common source of mistakes in management decision-making is the emphasis on finding the right answer rather than the right question. It is not enough to find the right answer; more important and more difficult is to make effective the course action decided upon. Management is not concerned with knowledge for its own sake; it is concerned with performance".

Conceptual Framework

Constraints

According to Merriam-Webster Dictionary Constraint means "the threat or use of force to prevent, restrict, or dictate the action or thought of others or the state of being restricted or confined within prescribed bounds; one, that restrict, limits or regulates". It is originated from Middle English, constreinte, from Old French, from feminine past participle of constraindre, to constrain. According to Rogers Thesaurus, constraint is something that restricts, check, circumscription, cramp, curb, inhibition, limit, limitation, restraint, restriction, stricture.

A number of constraints lie between the theory and full scale adoption. These constraints come in different categories, such as intellectual and knowledge, social, financial, technical, infrastructural and last but not the least policy (Friedrich and Kassam, 2009).

According to Wikipedia, the theory of constraints (TOC) is a management paradigm that views any manageable system as being limited in achieving more of its goals by a very small number of constraints. There is always at least one constraint, and TOC uses a focusing process to identify the constraint and restructure the rest of the organization around it.

A constraint is anything that prevents the system from achieving its goal. There are many ways that constraints can show up, but a core principle within TOC is that there are not tens or thousands of constraints. There is at least one but most only a few in any given system. Constraints can be internal or external to the system. Waghmare and Pandit (1982) defined 'constraints' in terms of causes hindering adoption of new technologies.

According to Kashern and Jones (1988), the term 'constraints' generally refers to barriers or impediments (technical, social, psychological and situational) confronted in achieving desired objectives. It can be thought of as negative force affecting the attainment of a desired goal.

The well-being of the rural population worldwide is invariably linked to the performance of the agricultural sector and to the sector's ability to cope with the challenges that result from rising population pressures, changing demand for food and agricultural products, resource scarcity, climate change, and greater production uncertainty. The World Development Report 2008 (World Bank 2007) emphasizes agricultural extension as an important development intervention (1) for increasing the growth potential of the agricultural sector in the light of rising demand- and supply-side pressures, and (2) for promoting sustainable, inclusive, and pro-poor agricultural and hence economic development. The call for agricultural extension services is made at a time when the underutilization of the productivity and growth potential of the agricultural sector for development poses a severe threat for achieving food security and for further reducing (rural) poverty.

The degree of institutional and economic development determines the scope and types of agricultural extension services and the ways in which these services are provided and financed (Anderson, 2007). Ideally, the design of the service provision and funding arrangements reflects the inverse relationship between the stage of economic development and the importance of extension for agricultural development and poverty reduction. The access to well-defined extension services is more important for economies in which agriculture is a major or declining source of economic growth (agriculture-based and transforming economies) but less important in economies in which agriculture is a minor source of economic growth (urbanized economies).

A prominent example for the group of transforming economies is India. The transition from an agriculture-based to a transforming economy was initiated by macroeconomic and non-agricultural reforms in the early 1990s, which triggered unprecedented high growth in nonagricultural (urban) sectors. At the same time, a weak, ineffective, and inefficient extension system and the consequent deficits in knowledge and technology development and dissemination constrained agricultural-sector growth, which in turn caused the share of agriculture in aggregate income to contract from approximately 31% in 1993 to about 19% in 2003–05. In addition to constraining agricultural sector growth and thus rural development, a weak and ineffective extension system increased the exposure of the agricultural sector to the effects of high population growth, shifts in product demand, natural resource constraints, climate change, and HIV/AIDS, among others (Birner et al., 2006; Anderson 2007; World Bank, 2007).

In order to meet these challenges, India's extension system has experienced major changes since the late 1990s in governance structures, capacity, organization and management, and advisory methods. The changes involve the decentralization of extension service provision to the local level, the adoption of pluralistic modes of extension service provision and financing, the use of participatory extension approaches, capacity training of farmers to express their demands, and capacity training of service providers to respond to the demands of farmers, among others (Rivera, Qamar, and van Crowder 2001; Birner et al., 2006; Birner and Anderson 2007; Anderson 2007). The reform initiatives reflect the view that improvements in agricultural productivity require demand-driven and farmer-accountable, need specific, purpose-specific, and target-specific extension services.

Birner, et al. (2006) argue that there is no single optimal or best model for providing need specific, purpose-specific and target-specific extension services. The ultimate choice of the agricultural extension approach depends on (1) the policy environment, (2) the capacity of potential service providers, (3) the type of farming systems and the market access of farm households, and (4) the nature of the local communities, including their ability to cooperate. Different agricultural extension approaches can work well for different sets of farm conditions. In order to use extension approaches that best fit a particular situation, the agricultural extension system has to be sufficiently flexible to accommodate the different options. To this end, the recent agricultural-sector reforms have been geared toward creating a demand driven, broad-based, and holistic agricultural extension system (Sulaiman and Hall 2002, 2004; India, Planning Commission, 2005). This has involved the design and introduction of a multitude of integrated measures that, on the demand side, enable service users to voice their needs and hold service providers accountable, and on the supply side, influence the capacity of service providers to respond to the needs of the extension service users (that is, the farmers).

STRATEGY FOR OPERATIONALISATION OF RECOMMENDATIONS

The support to State Extension Programmes for Extension Reforms launched in May, 2005 is under implementation in 268 districts across the country and is in take off stage, however certain constraints in implementation are observed which are as follows:

Major Constraints

- Lack of convergence in operationalisation of extension reforms
- Lack of provision for dedicated manpower at various levels
- Inadequacy of funds
- Lack of infrastructural support below district level
- Inadequate support for promotion of Farmers Organizations and their federation.

Strategy to Overcome the Constraints

1. Lack of Convergence in Operationalisation of Extension Reforms

Convergence is one of the key guiding principles considered in operationalizing extension reforms. It is evident from the initial experience that the convergence has not taken place at various levels satisfactorily. Hence, it is proposed all the central and the state government funds earmarked for extension activities in agriculture and allied activities should be routed through ATMA for convergence and synergy.

Further, the extension activities such as large scale demonstrations, exposure visits, trainings, activities of farmer field schools, farm schools and development of farmer scientists should be addressed through District level extension activities under Extension Reforms.

2. Lack of Provision for Dedicated Manpower at Various Levels

It is observed that almost all the positions associated with extension reforms are made on adhoc basis or given additional charges. As the extension reforms envisage radical changes in the process and institutional mechanism of the existing extension system, dedicated manpower is a necessary condition. Initial experience shows that, dedicated Project Directors contributed significantly to the success of ATMA whereas, the performance of Project Directors with additional charges was quite unimpressive. Hence, it is proposed to make financial provision under the scheme to appoint dedicated officers at various levels.

State level—Director, core faculty and supporting staff of SAMETI.

District level—Project Director, Dy. Project Director and supporting staff.

3. *Inadequacy of Funds*

It is found that the funds are quite inadequate for undertaking extension activities at State/District/block levels for operational expenses as well as manpower. Hence, adequate fund provision may be made to support extension activities and manpower.

4. *Lack of Infrastructural Support below District Level*

The NATP experience proves that Farm Information and Advisory Center (FIAC) is an effective platform of ATMA at block level for the interface among and between members of Block Technology Team (BTT) and Farmer Advisory Committee (FAC). The infrastructure like training hall, teaching aids, viz. OHP, TV, LCD, computer, internet connectivity and furniture found to be very useful for carrying out various extension activities at block level and below. The connectivity at FIAC will provide the market related information, weather forecast, inputs availability and other updated technological information to the farming community on day-to-day basis. FIAC will be grass root level nodal point for consultation for large number of farmers where large-scale digitization of agriculture information can be channelized.

There is no alternate infrastructure facility is available to BTT and FAC at Block level to access such information. The FIAC would become single window delivery mechanism for information on agriculture and allied sectors at block level and below. It is envisaged that FIAC will be a forum for delivery of technological options besides being facility for training and accessing agriculture content on the net.

At present under Extension Reforms only 10% of the blocks are provided with infrastructural support at FIAC level with limited funds only for connectivity. In order to make use of FIAC for training and other extension activities, it has to be strengthened with a training hall, furniture and training aids. Hence, it is proposed to establish FIACs in all the developmental blocks with adequate infrastructural support. The infrastructure also may be created on Private-Public Partnership mode. The fiscal incentives may be provided to the private sector in such initiatives. Operationalisation of FIACs on Public-Private Partnership may also avoid the future financial liability on the state government.

5. *Inadequate Support for Promotion of Farmers Organisations (FOs) and Farmer Led Extension*

The role of farmers' organization in promotion of farmer-to-farmer extension and market linkages is evident from many studies. However, the present guidelines and financial provisions are insufficient to cover even a small portion of farming community. Hence, there is a need for increasing the financial outlay substantially to promote Farmers Organizations and Farmer led extension through massive expansion of farm schools, farmer field schools and organizing and capacity building of farmers in Agriculture and allied sectors.

Promotion of Farmers Organizations and their federation includes three activities namely, (i) organizing farmers around commodities; (ii) Capacity Building of Farmers in Agriculture for Farmer led Extension (CAFE) and (iii) Accreditation and involvement of NGOs/Private Sector in implementation of Developmental Programs.

i. *Organizing Farmers around Commodities:* The experience from various spheres emphasizes the need to organize farmers for getting them the benefits both mutually within the community and in interaction with external agencies supporting the development process. It is proposed to organize the farmers around commodities and federated at block/district/state level. Linking these groups to economic activities assumes greater significance in the present context to sustain the developmental efforts.

The main objective of promoting the Farmers Organizations is to create a mechanism at the village level among the farmer members to empower them for their own problem solving. This would also help in providing techno-economic support to the groups, enhance scale of operation, improve performance, promote infrastructure, improve access to resources, technology and markets, build the capacity of farmers and ultimately improve the economy of the farmers. Emphasis would be given to involve women farmers in the process of development. The whole process will be oriented to develop the capacity of farmers to plan, draw support from all the related organizations to undertake technological, production, processing or marketing activities based on their needs and resources.

Farmers Organizations can effectively bring about a client driven agriculture research and extension system and can be an important mechanism in articulating specific research and extension needs accelerating technology dissemination and also in improving technical, managerial and marketing skills of member farmers. They enable the resource poor farmers the capacity to reach-up and pull down the research and extension services and exert influence to make the policies more relevant. FO's mobilize local resource and regulate their use to maintain a long-term base for productive activity and put available local resources to their most efficient use. Farmers Organizations exhibit better bargaining power in emerging market scenario like contract farming and future markets.

There are some successful cases of Farmer Organizations (FOs) acquiring advanced technology and professionalising agriculture. The Farmers Organizations in their collective endeavour make necessary arrangements for proper inputs supply, extension support, credit, collection of produce, processing and marketing in integrated manner to maximize returns on the investment. Establishment of a sustained linkage between the Farmers Organizations and agencies engaged in inputs and credit are most likely to reduce the cost of inputs, credit and storage. Similarly, establishment of linkages between producers through Farmers Organizations and the market would shorten the supply chain resulting in better realization for the farmers.

Commodity Interest Groups would be encouraged to function at Block, District and state levels. The federations may undertake additionally the activities of marketing, processing and market oriented services to Farmers Organizations. The members from these organizations on the various committees like Farmer Advisory committee at Block level, Governing Board at ATMA, State level coordination committees and other forums would reflect farmers need better.

It is better to promote at least one Commodity Interest Group per village. Based on the potential, another group will be also promoted on a different commodity/ enterprise in the same village. These groups would be promoted in a phased manner.

ii. *Capacity Building of Farmers in Agriculture for Farmer led Extension (CAFE):* One farmer from each Commodity Interest Group would be selected and provided with necessary capacity building support on end to end basis. The knowledge of the trained farmer would be available to other members of the group.

iii. *Involvement of NGOs/Private Sector in Implementation of Developmental Programs:* NGOs have potential to make significant contribution in implementation of extension activities with public funds. With a view to ensure quality implementation, steps need to be taken to identify and encourage NGOs having high performance. NGOs, agripreneurs, etc. need to be provided some service charge for implementation of extension activities with public funds.

Conceptual Framework

In order to analyze India's recent agricultural extension reforms, the authors apply the conceptual framework in Fig. 10.1, first presented in Birner and Palaniswamy (2006). The framework identifies the major governance structures, organizational and managerial characteristics, and frame conditions (e.g. socioeconomic characteristics) by which public-sector extension reforms can improve the organizational and managerial performance of service provision, lead to better public-sector governance outcomes, and generate sustainable pro-poor development. Public-sector governance outcomes can be evaluated in terms of the efficiency, effectiveness, and long-term sustainability of service provision, regulatory quality, rule of law, the degree of corruption, and equity aspects.

In this book, the agricultural extension reforms and their impacts are analyzed in terms of the underlying supply and demand components of extension service provision and related reform efforts. Demand-side approaches of public-sector service reforms aim at improving the ability of the private sector (such as farm households and profit-oriented firms) and the third sector (such as non-governmental organizations, farmers' organizations, and rural women's groups) to demand better governance and to hold public officials accountable by strengthening the voice of clients. To this end, demand-side approaches include policies that increase information and coordination in voting, strengthen the citizens' right to information, and improve the credibility of political promises. Demand-side approaches of rural service provision also involve policies that promote the political decentralization of service delivery to local governments, reserve seats in local councils for women, and advocate participatory planning and implementation methods, among others. Figure 10.1 indicates that demand-side reforms are likely to be more effective if they directly address socioeconomic and sociocultural obstacles that prevent citizens from exercising their voice and demanding accountability.

Strategies to strengthen the demand side of rural service provision will have a little effect, if they are not accompanied by strategies to increase the capacity of service providers to finance and deliver the respective services, to apply the rules of law and regulation, and to control corruption. Supply-side approaches to public service delivery reforms include the administrative and fiscal decentralization of service delivery, public expenditure management reforms, and training programs for public officials, changes in procurement and audit procedures, and efforts to coordinate the activities of government agencies and departments. Another popular supply-side approach reduces the tasks that are performed by public-sector agencies. The respective strategies include outsourcing of service provision to organizations in the private and third sectors, public–private partnerships, pluralistic forms of service delivery, devolution of authority to user groups, and privatization. Recent reform trends emphasize the need

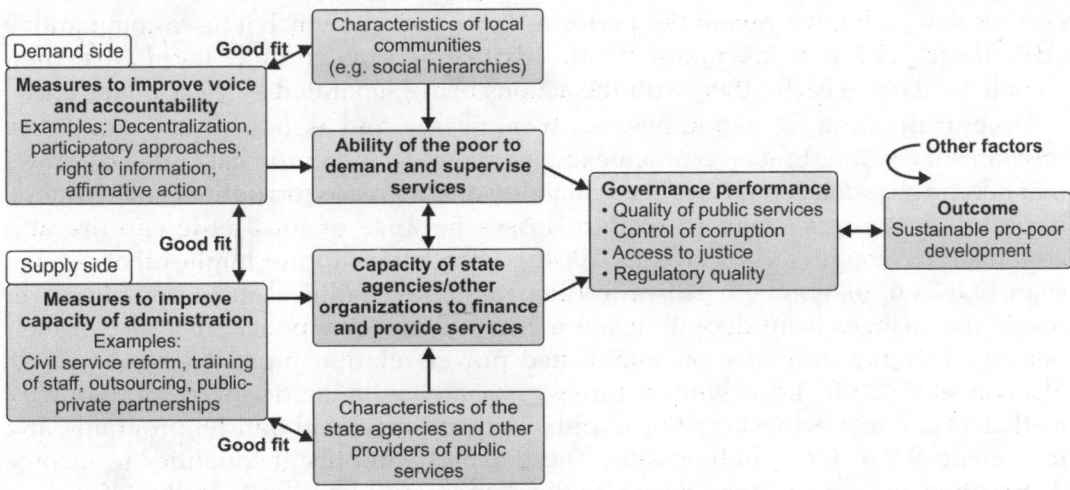

Fig. 10.1: Relationship among demand-side reforms, socioeconomic and sociocultural obstacles. *Source:* Birner and Palaniswamy (2006)

for the state to play a coordinating and facilitating role and to create an enabling environment for the private and third sectors. Supply-side approaches also include strategies for cost recovery that aim to improve the financial sustainability of service provision and to strengthen the incentive for clients to demand better services. The usefulness of cost recovery schemes can be debated on equity grounds, especially in absence of specific measures (e.g. vouchers for low-income households) that address this concern.

Figure 10.1 suggests that the success of demand- and supply-side reforms depends on the extent to which reform strategies address the sociocultural characteristics of local communities (e.g. social hierarchic structures) and the bureaucratic characteristics and incentive structures of public-sector service providers (such as moral and ethical standards and elite capture). Because local communities and service providers differ in terms of characteristics, a one-size-fits-all reform approach is an inadequate mechanism for improving rural service provision for rural development. In addition, the structure and scope of reforms also depends on the feasibility of reform implementation. Rather than engaging in ambitious reform programs that address all service delivery problems at the same time, it is often necessary to concentrate initially on those reform elements for which political support can be built (Levy, 2004). These relationships suggest that reform approaches should center on principles of best fit rather than best practice.

The conceptual framework in Fig. 10.1 points to the role of decentralization as a governance mechanism to improve the quality of and the access to basic services, infrastructure, and legal and regulatory structures. The importance of decentralization for good or responsive governance is attributable to the positive effect of decentralization on the efficient use of resources, with efficiency gains arising from (1) the functional, financial, and administrative autonomy of individual government tiers; (2) role clarity; (3) people's participation; (4) accountability; and (5) transparency, among others (see, e.g. von Braun and Grote 2002). At the same time, decentralization ensures that each government tier performs those tasks in which it has a comparative advantage. The actions taken by the different government levels are then complementary to each other, with the actions being separated by clear boundaries.

Decentralization is also subject to weaknesses and is not an accountability mechanism a priori. In fact, economic theory suggests and empirical evidence shows that decentralization can reduce accountability and increase corruption. Furthermore, decentralization may lead to welfare losses because of local elite capture and administrative failures (World Bank 2004b). Local elite capture implies that a small share of the population with a disproportionate share of political and economic power resists the changes from decentralization and participatory policies because of their perceived undue influence on established power relationships (Rajaraman, 2000; Narayan et al., 2000). Local elite capture is associated with inefficient use of resources, inefficient and ineffective targeting of public expenditures and transfer programs, and inefficient delivery of public goods. These aspects amplify inequalities in income distribution, with consequent threats to the economic and political stability of regions (Dethier, 2000; von Braun and Grote, 2002). Political and economic tensions may prevail not only within regions but also between them. In fact, they might be stronger if decentralization reinforces regional asymmetries in income distribution and income growth in the absence of cross-regional income compensation schemes.

SOME DEBATABLE ISSUES

1. Bottom up Planning Leads to Chaos and Conflict

Bottom-up planning is usually referred to as tactics. But in the area of resource-planning best practices, there has been considerable debate about whether top-down or bottom-up planning is "better". Support can be found for both approaches. Author Keith Duncan suggests that for organizations aiming to align limited resources with the most lucrative new product opportunities, a top-down approach provides the best balance of benefit to effort.

Figure 10.2 indicated that the top-down approach to integrated resource planning uses rough-order-of-magnitude sizing with very little detail to estimate resource needs. The bottom-up approach uses project planning techniques to create task-based estimates.

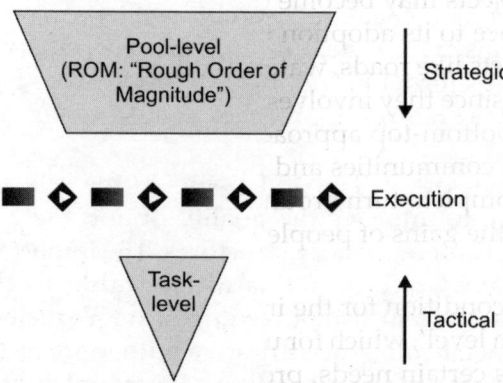

Fig. 10.2: Top-down vs bottom-up resource planning. *Source:* Keith Duncan (1996)

Bottom-top, otherwise referred as people's participation or democratic process. Participation in most cases varied according to the level of participation. It can be passive participation, Quasi or active participation. Participation has continued to metamorphose as modernization comes and one thing that is certain and constant in life is change. Change can be in either direction (positive or negative). In the words of Robert Chambers, "participation has implication for power relations, personal interactions, and attitudes and behaviours and that participatory can apply to almost all social contexts and processes, not least in organizations, education, research, communities and the family".

Cohen and Uphoff (1977) and Chambers (1993) were the early proposals of the theory and framework on participation. While Cohen and Uphoff (1977) had four dimensions of participation, viz. decision-making, implementation, benefits sharing and evaluation. Chambers (1993) had stages as project identification, prioritizing, planning, implementation, monitoring and evaluation. Hence the local communities are involved through consultation or by involving them in the partnership which makes them to see the programme as their own and put in all efforts to ensure the successful realization of the goals and sustainability of it too (Isidiho and Sabran, 2016).

According to Pissourios (2014), the adoption of bottom-top developmental approach may encounter the problem of legislation if the so called development was not covered in the existing central legislation. In this case it means the local communities has to work with the central law makers to enact laws to cover such rurally designed projects to guide against destruction and other illegalities and mitigations that might arise out of the projects. The laws would be able to spell out the do's and don't regarding the projects for its completion and sustainability. Also applicable might be the problems of full participation and control of the projects in larger communities if the bottom-top approach is adopted. This is because it may have to take longer time for the larger communities to assemble.

Debate and consensus on certain features, procedures and sharing of the programmes considering the fact that each participating unit would want to have preferential and greater influences in the location and operational modalities of the projects. Such may lead to serious conflict which at the end may lead to distrust, violence, inter and intra personality conflicts. The externalities impact and reflections of some bottom-top projects may become difficult for the communities to internalize hence may be a hindrance to its adoption in some developmental communities. Such big infrastructural projects like roads, water, electricity, airports may need to undergo top-bottom approaches since they involves much technicalities and external supports. Hence the adoption of bottom-top approaches depends on the projects involved and the strength of the local communities and their sponsors. The application of bottom-top approach is not a complete turn around on the top-bottom but a way of making amends and increasing the gains of people's active participation in the affairs of their life and well-being.

The fundamental precondition for the implementation of a bottom-up approach is the existence of a 'bottom level', which for urban planning corresponds to the existence of a community that has certain needs, problems and expectations, that are different from other communities, and is also willing to participate in planning procedures in order to influence them. However, on certain occasions there is no 'bottom level'. This may be the case in the planning of a new settlement or a large city plan expansion. On such planning occasions, there are not any residents yet, so the utilization of a bottom-up approach is attainable and planners can only turn to top-down approaches.

In the case that there is a local community which is willing to participate in planning processes, an assumption that is quite challengeable, the implementation of a bottom-up approach meets certain other obstacles. One of them is the relative difficulty in translating a bottom-up procedure into legislation. The existence of some sort of legislation is crucial, as it provides formalized rules and procedures that can maintain the agreement reached through the participative process (Healey, 1997). If any community is going to develop its own bottom-up planning process, i.e. a perspective to which the communicative approach adheres (Healey, 1997)—then inevitably the legislation should also be subject to the local community. However, their perspective on local lawmaking power is a far cry from the current administrative and constitutional organization of modern western states with a federal structure.

Participatory processes become cumbersome when the population size increases, slowing down the process of intervention, which is already a time-consuming process. In particular, the gathering of the various stakeholders of the community, the arrangement of the procedure in which the open-ended forms of discussions will be held, the arrival at the agreement or conflicted and interrelated issues and the translation of these agreements into planning objectives require the ampleness of time. Thus, in large communities, either the bottom-up processes will be inefficient, due to the slow progress of participatory processes, or techniques of representative participation will be adopted, which degenerate the nature of the bottom-up approach.

A third weakness of a bottom-up approach that further limits its scope is that such an approach can be implemented when planning deals with spatial issues as related to local interests and consequences only. Naess (2001) has argued on the weakness of bottom up approaches in the field of sustainable development, where a higher level of coordination is necessary. The same also applies for objectives that have consequences far beyond the local borders or their planning is affected by the preferences and needs of the residents outside these local borders. Such objectives are related to the location of supra local facilities.

It is obvious that bottom-up approaches are unable to guide either regional planning or strategic planning, as on the one hand such approaches cannot deal with the allocation and the location of supra local facilities and, on the other hand, their implementation becomes cumbersome, due to the large population size of the planned communities. Thus, for regional and strategic planning, a top-down approach is inevitably the only available choice for planning practice. As a result, the scope of bottom-up approaches is limited to the local planning of small settlements, or to the planning of districts in larger settlements. On these occasions of planning, the higher level strategic planning has already indicated the long-term objectives, for which the contribution of local participation is debatable, and has also resolved the conflict interest among neighboring settlements or districts. In addition, because of the small study area, participants are likely to have a clear and comprehensive view of their communities' strengths, weaknesses and opportunities, so their participation in planning procedure can be beneficial to the understanding of local needs, while participatory processes can be quick and flexible.

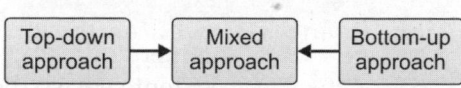

Fig. 10.3: Mixed approach. *Source:* Isidiho and Sabran (2016)

Both top-down and bottom-up approaches in community development are interwoven as there is no one that is perfect and encompasses developmental projects. There should be a blending or mixing of both to achieve a holistic and appreciable sustainable development that carries every one along. Hence the choice of mixed approach cuts across both top-down and bottom-top which shows in Fig. 10.3. The level of awareness in the twenty first century through the social media, open democracy and people oriented incorporation in governance and development calls for full integration of both approaches, making the author's proposed mixed approach stand the test of time as a new development paradigm.

2. Extension Work is More Important than Extension Education

Agricultural extension serves as a bridge between the agricultural research scientists and the potential users of research findings. It aims at helping research scientists to design and undertake needs and a problem based research and at the same time encourages and enables farmers and others to adopt new scientific knowledge and useful research results/agricultural technologies for increasing agricultural production leading to enhanced farm income and better quality of life for all in the rural area.

Extension Education at a Glance

1. It is a system in which youth both male and female, adult people and women are encouraged to work for their own development and prosperity.
2. Extension education is an informal system of education for all rural people.
3. Extension education brings required change among the people of rural area.
4. Extension education helps only those people who are prepared to help themselves.
5. It educates people as to how to achieve their target.
6. Extension education educates people through "learning by doing" and "seeing is believing".
7. Extension education is a two-way system of education. It brings scientific knowledge to the rural people and conveys the problems of rural people to the scientific institution for solution.
8. Extension education works together with the rural people, and thereby helps people to bring prosperity to their home, family, society, community and country.
9. Extension education helps in personal development, raising standard of living of people, developing local leaders and development of society and country.
10. Extension education is a continuous educational process, which goes on between teachers and learners. In this both teacher and learner benefit and learn from each other.

Extension Work

Extension work includes the process of extension education, i.e. the process of teaching and learning. Besides the process, in extension work are included organizations, administration, supervision, finances as well as the programs for the overall development.

The extension worker's role lies in mastering the extension technologies, educate villagers about their programmes get them interested in examining the new ways of thinking and help them try out innovation.

Extension Works with Rural People

Only the people themselves can make decisions about the way they will farm or live and an extension agent does not try to take these decisions for them. Rural people can and do make wise decisions about their problems if they are given full information including possible alternative solutions. By making decisions, people gain self confidence. Extension, therefore, presents facts, helps people to solve problems and encourages farmers to make decisions. People have more confidence in programmes and decisions which they have made themselves than in those which are imposed upon them.

Agricultural innovation systems are a paradigm in which information dissemination need not be performed only by extension workers. Extension workers must have an extensive knowledge of various agricultural disciplines and they should have the ability to deal with farmers and persuade them to adopt modern agricultural techniques and ideas so as to use them on their farms (Robertson, 2012).

Principle of Extension Work

Adaptability principles in use of extension teaching methods: Extension worker should have knowledge of extension methods so that they can select proper method according to the condition. Teaching method should be flexible so that they can be properly applied on people according to their age groups, educational background, economic standard and gender. In extension education two or more methods should be applied according to principle of adaptability.

Principle of leadership: The participation or inclusion of local leaders in extension programmes is the only criteria for assessing the success or failure of any extension work.

Principle of satisfaction: If people are not interested in extension work then there is no possibility that extension work can be carried on for a long-time. In democratic structure/set-up people cannot be run in a mechanized way. They should derive full satisfaction from extension work. Extension worker should give priority to those work in which there is scope for immediate benefits. Primary satisfaction is very helpful for the future of extension work.

Principle of evaluation: It is necessary to evaluate the extension work after a certain period so that merits and demerits of extension work can come to light and necessary changes be brought about. Evaluation generates confidence in people.

Principle of neutrality: Extension worker should never take interest in local politics. If he will not behave in this manner then lots of difficulties in extension work will arise. Therefore, remaining neutral is more beneficial. Extension worker should never express his special affection or hatred towards any person.

Principle of encouragement: In extension work, principles of encouragement have great importance. Under pressure no work can be done in extension programme, for this active workers in this field should be encouraged so that they participate and enthusiastically remain active.

Principle of trained specialists: It is very difficult that extension personnel should be knowledgeable about all problems. Therefore, it is necessary that specialists should impart training to the farmers from time to time.

Role of Extension Worker

Extension worker is like a teacher, philosopher, leader, guide and colleague for the community and on the basis of his special qualities he is able to carry on developmental programme successfully. It is undoubtedly true to say that the success and failure of development programmes depend upon the qualities of extension worker. The extension worker's role is to teach the farmers about how to use new technologies. Knowledge and application of extension education principles, the extension workers help a lot in determining the needs constraints, priorities and opportunities for farmers. They also help in teaching farmers the value of improved agriculture, recommending suitable crops, encouraging adopting of appropriate technologies, and evaluating farmers' reaction and attitudes toward development projects.

The role of extension works in agricultural technology transfer (Aremu et al., 2015):

- Extension worker create awareness of innovation, something new or perceived to be unknown before to his clients.
- Persuasion of the usefulness or importance of the new technology.
- Reinforcement of continued use of technology that is created. The interest to continue to practice the new idea that was introduced is sustained through supporting services like input.
- Intermediary between the researchers and farmers. This is done by creating awareness to the farmers, and also taking of farmer's problem to the researchers.
- Diagnose problems by telling them the way out.
- Extension workers stabilize change and attempts to prevent discontinuance, individuals tend to seek confirming information for the decision they may make.
- Extension workers can effectively stabilize new behavior by directing, and reinforcing messages to those clients who have adopted innovation.
- Extension workers act as catalyst to speed up the rate at which his clients accept changes or innovation.
- Extension worker is a solution provider that is he or she has an idea about effective solution to farmers problems.
- Extension worker play a role of resource linker that is linking them with necessary agricultural inputs, to locate source of finance for their farms.
- Strengthening and supporting farmer organization.

Agricultural extension services are very important in the development of rural knowledge and innovative systems for farmers. These services are key in informing and influencing rural household decisions, especially in the developing countries which are generally more in need for such guidance services (Alex et al., 2002).

The duties of agricultural extension workers include introducing farmers at agricultural training courses into training program on various agricultural subjects in order to provide them with information and knowledge about methods and techniques of agriculture, and, consequently, increase their production and income efficiency, improve their living, and raise the social and educational standards of rural life. Agricultural extension takes care in the youth and women in rural areas and enables them to develop their knowledge of various subjects concerning agricultural and social issues. Agricultural extension focuses on two main facets, as follows (Bello et al., 2008).

- It deals with the behavior of rural people in terms of influencing them through education and the exchange of information. The aim is to assist the people in gaining a livelihood, improving the physical and psychological level of living of rural families and fostering rural community welfare.

- As a service, agricultural extension makes a government ministry, a university or a voluntary agency as useful as possible to the people who support it through taxes and donations.

Knowledge is a prerequisite and therefore a qualification in at least one field of technical agriculture is a necessity for the extension worker. If the extension worker wants to be successful she/he must be able to communicate in agriculture. What is expected today is that every extension agent must be an expert in at least one field of technical agriculture. To be successful the extension agent must be technically empowered in agriculture to deliver a service of excellence. The extension agent must have the knowledge and skills to plan a farm physically, biologically and economically, as well as the skills to adapt and transform the technical message to be applicable and sustainable to the specific farm and farmer.

Completing a successful extension educational program is very fulfilling and rewarding experience. In fact, it is one of the greatest feelings of accomplishment that

Table 10.1: Extension education versus extension work

Extension education	Extension work
Extension education starts with theory and works up to practical	The extension workers starts with practical and may take up theory later on
Students and extension worker study subject	Farmers study problems
Fixed and flexible curriculum offered	It has no fixed curriculum or course of study and the farmers help to formulate the curriculum
Teacher instructs the students, extension worker	Teacher teaches and also learns from the farmers
Extension education is mainly vertical and horizontal	Extension work is mainly horizontal
Extension educationist more or less homogeneous audience	The extension worker has a large and heterogeneous audience
It is rigid as well as flexible	It is flexible
It has all pre-planned and pre-decided programmes	It has freedom to develop programmes locally and they are based on the needs and expresses desires of the people
It is more theoretical	It is more practical and intended for immediate application in the solution of problems

occurs through extension work. Developing and delivering an effective extension educational program can be a challenge, and it requires a great deal of commitment. For some extension professionals, this may be an aspect of the job that they find most difficult. It may be a new challenge for faculty who have had little education, knowledge or experience in this area. How one carries out the various phases of an educational program directly affects its success and eventual outcomes.

It may concluded from the above study that work of agricultural extension is the basis for the development of the agricultural sector, and without agricultural extension, it does not have any benefit from modern agricultural techniques and modern agricultural information. Agricultural extension work is responsible for the transfer of agricultural technologies to farmers, and to convince farmers to adopt modern agricultural techniques. Agricultural extension work is the bridge that connects farmers with agricultural research centers in order to transfer all agricultural techniques to farmers and teach them how to use them in their farms. It is also concluded that the agricultural extension workers have an effective and important role in helping farmers solve agricultural problems.

Extension education is the process of arranging situations in which the things to be learnt are brought to the notice of the learners, their interest is developed and desire aroused, i.e. they are stimulated to action. The extension–teaching methods are the tools and techniques used to create situations in which communication can take place between the rural people and the extension workers. They are the methods of extending new knowledge and skills to the rural people by drawing their attention towards them, arousing their interest and helping them to have a successful experience of the new practice. A proper understanding of these methods and their selection for a particular type of work are necessary. So without proper knowledge of extension education an extension worker unable to reach an organization objective and it is also clear from the study that value of extension education is no more without doing any extension work. So there is need for application of extension education through extension work in farming community. So extension work is more important than extension education for development of socioeconomic condition of people and developing and transforming new and existing agricultural knowledge.

Bibliography

1. Agriclinics and Agribusiness Centres (2016). Progress of Agriclinics and Agribusiness Centres Scheme, Jan 8, available at http://www.agriclinics.net/state-wise.pdf, viewed on January 16, 2016.

2. Agricultural Knowledge Management: Global Extension Experiences from November 9-12, 2011 at NAAS New Delhi organized by International Society of Extension Education,

3. Alex G, Zijp W, Byerlee D. (2002). Rural Extension and Advisory Services: New Directions, with Input from the AKIS Thematic Team. *Rural Development Strategy Background* Paper No.9. Washington DC. The World Bank.

4. Alex G, Zijp W (2002). Rural extension and advisory services. Rural development strategy background, World Bank, USA.

5. Al-Rimawi A, Al-Karablieh E (2002). Agricultural Private Firms' Willingness to Cooperate with Public Research and Extension in Jordan. *Journal of International Agricultural and Extension Education*, 9, 5.

6. Anandajayasekeram P (2008). Concepts and practices in agricultural extension in developing countries: A source book. ILRI (aka ILCA and ILRAD).

7. Anandajayasekeram P, Purkur R, Sindu W, Hockstra D (2008). *Concepts and practices in agricultural extension in developing countries: A source book*. Washington DC, USA and Nairobi, Kenya. IFPRI and ILRI.

8. Anderson JR. (2007). Agricultural advisory services. Background paper for the World Development Report 2008. Agriculture and Rural Development Department. Washington DC: World Bank.

9. Angstreich M, Zinnah M. A meeting of the minds: farmer, extensionist, and researcher. *Journal of International Agricultural and Extension Education*, 2007;14:85–95.

10. Aremu P, Kolo I, Gana A, Adelere F. "The critical role of extension workers in agricultural technologies transfer and adoption". *Global advanced research Journal of food science and technology*, 2005;4(2):14–8.

11. Asenso-Okyere K, Davis K, Aredo D (2008). Advancing agriculture in developing countries through knowledge and innovation: Synopsis of an international conference. *Intl Food Policy Res Inst.*

12. Barghouti S, Kane S, Sorby K, Ali M, (2004). "Agricultural diversification for the poor: Guidelines for practitioners", *Agriculture and Rural Development discussion paper*, World Bank, Washington, DC.

13. Barrick RK, Samy MM, Gunderson MA, Thoron AC. A model for developing a well prepared agricultural workforce in an international setting. *Journal of International Agricultural and Extension Education*, 2009;16(3):25–31.

14. Bello A. (2008). Introduction to agricultural extension and rural sociology. National Open University of Nigeria, Nigeria.

15. Bernardo FA (1986). Report of the Consultation Meeting of Agricultural Research and Extension Experts, SEARCA.

16. Bharati RC, Singh KM, Chandra, Singh AK. Economic condition of eastern region of India-An statistical evaluation. *Journal of AgriSearch* 2014;1(3):173–9.

17. Bingen J, Dembèlé E. 2004. Mali: The Business of Extension Reform— Cotton in Mali. In W. Rivera and G. Alex, eds., *Extension Reform for Rural Development* (vol. 2, pp. 83–7). Washington, DC: World Bank. (http://www-wds .worldbank.org/external/default/ WDSContentServer/WDSP/IB/2005/04/05/ 000090341_20050405102736/Rendered/PDF/ 318910Extension1Reform1V21final.pdf)

18. Birner R, Anderson J (2007). How to make agricultural extension demand-driven? The case of India's agricultural extension policy. Discussion Paper 00729. Washington D.C.: International Food Policy Research Institute.

19. Birner R, Palaniswamy N. (2006). *Public administration reform and rural service provision: A comparison of India and China.* Manuscript. Washington D.C.: International Food Policy Research Institute.

20. Birner R, Davis K, Pender J, Nkonya E, Anandajayasekeram P, Ekboir J, Cohen M. From best practice to best fit: a framework for designing and analyzing pluralistic agricultural advisory services worldwide. *Journal of Agricultural Education and Extension*, 2009;15(4): 341–55.

21. Birner RK, Davis J, Pender E, Nkonya P, Anandajayasekeram J, Ekboir A, Mbabu DJ, Spielman D, Horna S, Benin, M Cohen. (2006). From best practice to best fit: A framework for analyzing pluralistic agricultural advisory services worldwide. Development Strategy and Governance Division Discussion Paper No. 37. Washington DC: International Food Policy Research Institute.

22. Boundless (2014) "The step of Purchase decision", Retrieved from https://www.boundless. Com/marketing/consumer-marketing/consumer-decision process/purchase.

23. Bozman B. "Technology transfer and public policy: a review of research and theory". *Research Policy*, Elsevier, 2000;29(4-5):627–55.

24. Braun von J, Grote U (2002). Does decentralization serve the poor? In *Managing fiscal decentralization,* ed. E. Ahmad and V. Tanzi. London and New York: Routledge-Taylor and Francis Group, Routledge Economics.

25. Brewer F (2000). *History of Indian Extension.*

26. Byrnes KJ (2001). Farmer Organizations: Tapping their Potential as Catalysts for Change in Small- Farmer Agricultural Systems. In F. Byrnes (ed.) Agricultural Extension System: An International Perspective. North Chelmsford, MA: Erudition Books.

27. Chambers R (1983). Rural Development: Putting the Last First. London: Longman

28. Chambers R (1993). Challenging the Profession-Frontier for Rural Development, London: *Intermediate Technology Publications.*

29. Chambers R (1997). Whose Reality Counts? *London Intermediate Technology Publications.*

30. Chambers R, Jiggins J (1986). Agricultural Research for Resource Poor Farmers. *IDS Discussion paper 220.* Brighton: University of Sussex.

31. Chambers RG. Information, incentives, and the design of agricultural policies. *Handbook of Agricultural Economics*, 2002;2:1751–1825.

32. Chambers, Robert (1993). Challenging the professions: frontiers for rural development, London: IT Publications.

33. Chandrasekhar Rao N, Srinivasan J, Das Gupta S, Reardon T, Minten B, Mehta M (2011). Agri-services in Andra Pradesh for Inclusive Rural Growth: Baseline Survey Findings and Policy Implications. Submitted to USAID, New Delhi. International Food Policy Research Institute. New Delhi.

34. Chen M (1996). Managing International Technology Transfer. Thunderbird Series in International Management. London: International Thompson Press.

35. Choi HJ (2009). Technology Transfer Issues and a New Technology Transfer Model. *The Journal of Technology Studies,* Fall 2009, 35 (1).

36. Christoplos J, Nitsch U. Changing Extension Paradigm. *IDRC Currents*, 1993;6:22–6.

37. Chukwuone N, Agwu A, Ozor N (2006). Constraints and strategies toward effective cost-sharing of agricultural technology delivery in Nigeria: perception of farmers and agricultural extension personnel. *Journal of International Agricultural and Extension Education*, 13.

38. Cohen John M, Norman T Uphoff (1977). Rural Development Participation: Concept and Measures for Project Design, Implementation and Evaluation. Rural Development Committee, Cornell University.

39. Conway GR, Barbier EB. (1990). After the Green Revolution: Sustainable Agriculture for Development, London: *Earth Science Publication Ltd*.

40. Cooksey, Brian, Idris Kikula (2005). When Bottom-Up Meets Top-Down: The Limits Of Local Participation In Local Government Planning In Tanzania. Special Paper No: 17. Research on Poverty Alleviation. *Mkuki Na Nyota Publishers*. Tanzania.

41. Cousins JH, Rubovits DS. Adolescent risk-taking: An analysis of problem behaviors in problem children. *Journal of Experimental Child Psychology*, 1993;55(2):277–94.

42. Cummings FH. Role of participation in the evaluation and implementation of development projects. *Knowledge and Policy*, 1997;10(1-2):24–33.

43. DAC. (2000). Policy framework for agricultural extension. Department of Agriculture and Cooperation, Extension Division. Available at: http://agricoop.nic.in/policy_framework.htm, accessed on June 16, 2008.

44. DAC-DARE (2011). Convergence between Research and Extension. Department of Agriculture and Cooperation and Department of Agricultural Research and Education. Ministry of Agriculture, Government of India, New Delhi.

45. Davis K. Extension in sub-Saharan Africa: Overview and assessment of past and current models, and future prospects. *Journal of International Agricultural and Extension Education*, 2008;15:15–28.

46. Davis KE, Nkonya D, Ayalew, Kato E. *Assessing the Impact of a Farmer Field Schools Project in East Africa*. Paper presented at 25th Annual Conference of the Association for International Agricultural and Extension Education, San Juan, Puerto Rico, 2009;May 24–8.

47. Desai BM, D'souza E, Mellor JW, Sharma VP, Tamboli P. Agricultural policy strategy, instruments and implementation: A review and the road ahead. *Economic and Political Weekly*, 2011;42–50.

48. Dethier JJ. (2000). *Some remarks on fiscal decentralization and governance*. Paper prepared for presentation at the Conference on Decentralization Sequencing, Jakarta, Indonesia, March 20, 2000.

49. DeVore PW. (1987). Technology and Science. In E.N. Israel and R.T. Wright (Eds.), *Conducting Technical Research* (pp. 27-45). Mission Hills, CA: Glencoe.

50. Drucker PF. (1966). The Effective Executive. New York: Harper and Row.

51. Duncan KT. (1996). Linking Operations to Strategy and Tactics in the Dardanelles. NAVAL WAR COLL NEWPORT RI.

52. Durkheim E (1893). The Division of Labor, New York

53. Duvel G (1996). Institutional Linkages for Effective Coordination And Cooperation In Participative Rural Development. *Journal of International Agricultural and Extension Education*, 3, 76.

54. Ernest L. Okorley, David Gray, Janet Reid. Towards a Cross-Sector Pluralistic Agricultural Extension System in A Decentralized Policy Context: A Ghanaian Case Study. *Journal of Sustainable Development in Africa*, 2010;12(4), ISSN: 1520–5509.

55. FAO. (2001). The state of food insecurity in the world, 2001. Rome.

56. Food and Agriculture Organization (FAO) (2006a). The State of Food Insecurity in the World 2006. Rome: Food and Agriculture Organization of the United Nations. (www.fao.org/docrep/009/a0750e/a0750e00.htm).

57. Franklin Ursula. (1992). The Real World of Technology (CBC Massey lectures series.) Concord, ON: *House of Anansi Press Limited*. ISBN: 0-88784-531-2.

58. Frey RE (1987). Is There a Philosophy of Technology? *Paper presented at the 74th Mississippi Valley Industrial Teacher Education Conference*, Chicago, IL.

59. Friedrich T, Kassam AH (2009, February). Adoption of conservation agriculture technologies: constraints and opportunities. *In Invited paper, IV World Congress on Conservation Agriculture* (pp. 4–7).

60. Furby L, Beyth-Marom R. Risk taking in adolescence: A decision-making perspective. *Developmental review*, 1992;12(1):1–44.

61. Galbraith JK. (1967). The New Industrial State, Boston, MA: Houghton Mifflin.

62. Gibson DV, Rogers EM (1994). Research and Development on Trial: The Microelectronics and Computer Technology Consortium. Boston: *Harvard Business School Press*.

63. Glendenning CJ, Babu SC (2011). Decentralization of Public-Sector Agricultural Extension in India.

64. Glendenning CJ, Babu S, Asenso-Okyere K. (2010). Review of agricultural extension in India. Are farmers information needs being met. *Discussion Paper 01048*. Washington DC: International Food Policy Research Institute.

65. Grabowski R. The New Technology and Reform, *Indian Journal of Agricultural Economics*, 1967;42(5):52.

66. Hall A, Sulaiman VR, Berkorowajnyj P. (2008). Reframing Technical Change: Livestock Fodder Scarcity Revisited as innovation capacity scarcity. Nairobi, Kenya and Maastricht, Netherlands: ILRI and UNU/MERIT.

67. Hall AJ, Norman C, Taylor S, Sulaiman VR. (2001). Institutional Learning Through Technical Projects: Horticulture Technology Research and Development System in India. *Agriculture Research and Extension Network paper III*, London, Overseas Development Institute.

68. Hameri AP. Technology transfer between basic research and industry, *Technovation*, 1996;16(2):51–7.

69. Hashim MY (1978). The Development and Transfer of New or Improved Technology and Implication for Extension Training. In S Radhakrishna and MM Singh, *Technology for Rural Development (COSTED)*, Bangalore, Indian Institute of Science. http://web3.canr.msu.edu/vanburen/India/histupd.htm.

70. HEALEY P, 'Planning Through Debate: The Communicative Turn in Planning Theory', [in:] Campell S and Fainstein S (eds.), *Readings in Planning Theory, Massachusetts: Blackwell Publishers (originally published in 1992 in Town Planning Review*, 1996;63(2): 143–62.

71. Healey P (1997). *Collaborative Planning. Shaping Places in Fragmented Societies*, Vancouver: UBC Press.

72. Healey P, Mcdougall G, Thomas M (1982). 'Theoretical Debates in Planning Towards a Coherent Dialogue', [in:] HEALEY, P., MCDOUGALL, G. and THOMAS, M. (eds.), *Planning Theory. Prospects for the 1980s*, Oxford: Pergamon Press.

73. ICAR (2011). Vision 2030, New Delhi: Indian Council of Agricultural Research.

74. Idiho AO, Sabran MSB. An evaluation of the effectiveness of leadership in project implementation, governance and community development. *Australian International journal of Humanities and Social Studies*. 2015;2(11):11–22.

75. India, Planning Commission. (2005). *Midterm appraisal of the 10th Five Year Plan (2002-2007)*. Available at: http://planningcommission.nic.in/midterm/english-pdf/section -05.pdf, accessed on May 4, 2007.

76. International Society of Extension Education, Nagpur and Indian Council of Agricultural. Javier EQ 1989. Recent approaches in the study and management of the linkages between agricultural research and extension. *ISNAR Staff Notes* No. 89–63.

77. Johnson SD, Gatz EF, Hicks D. (1997). Expanding the content base of technology education: Technology transfer as a topic of study. Volume 8 Issue 2 (spring 1997).

78. Jones GF, Garforth C (1997). The history, development, and future of agricultural extension. In B. E. Swanson, R. P. Bentz, & A. J. Sofranko (Eds.), Improving agricultural extension (p. 220). Retrieved from http://www.fao.org/docrep/W5830E/w5830e03.htm.

79. Kashern MA, Jones GF. (1988). Obstacles in individual innovation decision making. *Indian Journal of Extension Education.* XXIV (3 -4), 2-8.

80. Katz E. (2002). Innovative Approaches to Financing Extension for Agricultural and Natural Resource Management. Landau, Switzerland; LLBL, Swiss Centre for Agricultural Extension.

81. Keizer Piet (2008).The Concept of Institution: Context and Meaning. Utrecht University School of Economics, Utrecht University, Janskerkhof 12, 3512 BL Utrecht.

82. Kim J, Kong M, Ju D. Challenges in Public Agricultural Extension of Korea. *Journal of International Agricultural and Extension Education*, 2009;16.

83. Kolfer VL, Meshkati N. (1987). Transfer of Technology: Factors for Success, In M.J. Marquardt (Ed.), Corporate Culture: International HRD Perspectives. (pp. 70–85). *Alexandria, VA*: American Society for Training and Development.

84. Kristin Davis, Place NT (2003). Non-governmental Organizations as an Important Actor in Agricultural Extension in a Semiarid East Africa. *Journal of International Agricultural and Extension Education*, 10.

85. Kumar B, Hansra BS (2001). Extension Education for Human Resource Development. New Delhi: *Concept Publishing Company.*

86. Levy B. (2004). Governance and economic development in Africa: Meeting the challenge of capacity building. In *Building state capacity in Africa*, ed. B. Levy and S. Kpundeh. Washington, D.C. and London: World Bank and Oxford University Press.

87. López J, Bruening T. Meeting Educational Needs of San Lázaro Farmers: Indigenous Knowledge Systems. *Journal of International Agricultural and Extension Education*, 2002;9: 39.

88. Lundquist G. A rich vision of technology transfer technology value management. *The Journal of Technology Transfer*, 2003;28(3-4):265–84.

89. MANAGE (1999). Innovation in Technology Dissemination. NATP Series-1. National Institute of Agricultural Extension Management, Hyderabad, India. MANAGE.

90. Mansfield E. International technology transfer: forms, resource requirements, and policies. *The American Economic Review*, 1975;65(2):372–6.

91. Markert LR. (1993). Contemporary Technology: Innovations, Issues and Perspectives. (Chapter. 8, pp. 231–253). South Holland II: Goodheart Willcox.

92. McDermott JK. (1987). Making extension effective: the role of extension/research linkages. *in:* Rivera W, and Schram S (eds) 1987. *Agricultural Research Worldwide.* New York, NY: Croom Helm.

93. Meena MS, Singh KM, Swanson B. Indian Agricultural Extension Systems and Lessons Learnt: A Review. *Journal of AgriSearch* 2015;2(4):281–5.

94. Meena M, Singh KM (2011). Measurement of attitudes of rural poor towards SHGs in Bihar, India. *India* (Nov 1, 2011).

95. Meena MS, Bhagwat VR, Jain RK, Ilyas SM. (2003). Self-Help Group: An Effective Instrument for Transfer of Technology. CIPHET Extension Bulletin-1.Central Institute of Post-Harvest Engineering & Technology, Ludhiana, India.

96. Meena MS, Jain D, Meen HR. Measurement of Attitudes of Rural Women towards Self-Help Groups, *The Journal of Agricultural Education and Extension* 2008;14(3):217–29.

97. Mitcham, C. (1980). Philosophy of technology. In Durbin, P. T. (Ed.), *A Guide to the Culture of Science, Technology and Medicine*. (pp. 282–363). New York: Free Press.

98. Mittelman JH, Pasha MK. (2016). Out from underdevelopment revisited: Changing global structures and the remaking of the Third World. Springer.

99. Moris J. (1991). Extension Alternatives in Tropical Africa. London: Overseas Development Institute.

100. Mwabu G, Thorbecke E. Rural Development Economic Growth and Poverty Reduction in Sub-Saharan Africa. *Journal of African Economics*. 2001;13:116–65.

101. Mytelka LR. Local Systems of Innovation in a Globalised World Economy. *Industry and Innovation*. 2000;7(1):16–32.

102. Naess P. 'Urban Planning and Sustainable Development', *European Planning Studies,* 2001;9 (4):503–24.

103. Nagel UJ (1997). Alternative approaches to organizing extension. Improving agricultural extension.

104. Narayan D, Patel RK, Schafft A. Rademacher, Koch-Schulte, S. (2000). *Voices of the poor. Can anyone hear us?* Washington DC: Oxford University Press and World Bank.

105. Narayan N (1993). Participatory Evaluation Tools for Managing Change in Water and Sanitation. *World Bank Technical Paper 207*. Washington DC, USA: The World Bank.

106. National Research Council. (1987). Management of Technology: The Hidden Competitive Advantage. Washington, DC: *The National Academies Press*. https://doi.org/10.17226/18890.

107. National Sample Survey Organisation (NSSO). (2014). Key Indicators of Situation of Agricultural Households in India, 70th Round (January–December 2013), Ministry of Statistics and Programme Implementation, New Delhi.

108. Nayak MP (2015). "Management of Institutional Reforms and its' impact under Innovation in Technology Dissemination (ITD) component of National Agricultural Technology Project (NATP): A study in the state of Orissa", a Ph. D Thesis, Department of EES, PSB, Visva-Bharati, (unpublished).

109. Niamh Dennehy, Dermot J Ruane, Phelan JF. Supports and Funding for Community Development Projects in the Republic of Ireland. *Journal of International Agricultural and Extension Education*, 2000;7:83.

110. Nigam SN, Gowda CLL (1996). Technology Development and Transfer in Agriculture in Achieving High Groundnut Yields. *Preceeding of an International workshop*. (pp. 183–93). 25–29 Aug, 1995, Laixi City, Shandong, China.

111. NSSO (National Sample Survey Organization). (2005). Situation assessment survey of farmers: Access to modern technology for farming, 59th round (January–December 2003). Report No. 499(59/33/2). New Delhi: Ministry of Statistics and Programme Implementation.

112. Oakley P. The concept of participation in development. *Landscape and Urban Planning,* 1991;20(1-3):115–22.

113. Okorley E, Gray D, Reid J. Improving Agricultural Extension Human Resource Capacity in a Decentralized Policy Context: A Ghanaian Case Study. *Journal of International Agricultural and Extension Education*. 2009;2:38.

114. Osman-Gani AAM. (1999). International technology transfer for competitive advantage: A conceptual analysis of the role of HRD. *Competitiveness Review*, 9.

115. Owona Ndongo P, Nyaka Ngobisa A, Ehabe E, Chambon B, Bruneau J. Assessment of training needs of rubber farmers in the South-west region of Cameroon. 2010;5:2326–31.

116. Parsai G (2010). Double Farm Growth Rate to Ensure Food Security. *The Hindu*, June 19, 2010.

117. Parvan A (2011). Agricultural Technology Adoption: Issues for Consideration when Scaling-Up. *Cornell Institute for Public Affairs*. Cornell University.

118. Phillips RG. Technology business incubators: how effective as technology transfer mechanisms? *Technology in Society*, 2002;24(3):299–316.

119. Pissourios IA. 'Whither the Planning Theory–Practice Gap? A Case Study on the Relationship between Urban Indicators and Planning Theories', Theoretical and Empirical Researches in Urban Management, 2013b;8 (2):80–92.

120. Planning Commission. (2002). 10th Five Year Plan 2002-07: Sectoral Policies and Programmes, New Delhi, India: Planning Commission. *Yojana Bhavan*, Govt. of India.

121. Planning Commission. (2007). Recommendation of Working Group on Agricultural Extension for Formulation of 11th Five Year Plan (2007-12), New Delhi, India: Planning Commission. *Yojana Bhavan*, Govt. of India.

122. Planning Commission (2008). 11th Five Year Plan (2007–12). Ministry of Finance, Government of India, New Delhi.

123. Porter G (2001). Accountability, Subsidiarity and Diversity. *Retrieved on March*, 5.

124. Prasad C. Agricultural extension services. In 40 Years of Agricultural Research and Education in India, 1989;234–67. New Delhi, India: ICAR.

125. Qamar MK. Global Trends in Agricultural Extension: Challenges facing Asia and Pacific Region. In FAO regional expert consultation on agricultural extension, research-extension-farmer interface and technology transfer. *Proceeding of consultation held July, 2002;16–19 Bangkok, Thailand*. Bangkok: FAO Regional Office for Asia and the Pacific.

126. Raabe K (2008). Reforming the Agricultural Extension System in India: What Do We Know About What Works Where and Why? *Discussion Paper 00775*. Washington DC: International Food Policy Research Institute.

127. Rajalahti R, Janssen W, Pehu E. (2008). Agricultural innovation systems: From diagnostics toward operational practices. Agriculture & Rural Development Department, World Bank.

128. Rajaraman I (2000). Fiscal features of local government in India. In *Governance, decentralization, and reform in China, India, and Russia*, ed. JJ Dethier. Dordrecht: Kluwer Academic Press.

129. Ramanathan K. The polytrophic components of manufacturing technology. *Technological Forecasting and Social Change*, 1994;46:221–58.

130. Ramanathan KA. Evaluating the national science and technology base: A case study in Sri Lanka. *Science and Public Policy*, 1989;15:304–20.

131. Reardon T, Minten M, Das Gupta S, Singh S. Synthesis- Agri-services in Uttar Pradesh for Inclusive Rural Growth: Baseline Survey Findings and Policy Implications. Submitted May, 2011 to USAID, New Delhi. International Food Policy Research Institute. New Delhi.

132. Reddy MN, Swanson B. *Strategy for up-scaling the ATMA model in India*. Annual Conference Proceedings of the Association for International Agricultural and Extension Education 2006;22:561–9.

133. Rivera, et al. (2005). Agricultural extension as component of an agricultural knowledge and innovation system. In Anandajayasekeram, P., Puskar, R., Sinda, W. and Hoekstra, D. (2008). Concepts and practices in agricultural extension in developing countries: A source book. (pp.275). Washington DC, USA and Nairobi, Kenya. IFPRI AND ILRI.

134. Rivera W (2001). Agricultural and rural extension worldwide: options for institutional reform in the developing countries. Extension, Education and Communication Services, Research, Extension and Training Division. *Sustainable Development Department*, Rome: FAO.

135. Rivera W, Alex G. The continuing role of government in pluralistic extension systems. *Journal of International Agricultural and Extension Education*, 2004;11:41–52.

136. Rivera WM, Alex G, Hanson J, Birner R (2006). Enabling agriculture: The evolution and promise of agricultural knowledge frameworks. In *Proceedings of the Association for International Agricultural and Extension Education Annual Conference*, Clearwater Beach, FL.

137. Rivera WM, Elshafie EM, Aboul-Seoud KH. The Public Sector Agricultural Extension System in Egypt: A Pluralistic Complex in Transition, *Journal of International agricultural and Extension Education*, 1997;4(3):56.

138. Rivera W, Qamar MK, L van Crowder. (2001). Agricultural and rural extension worldwide: Options for institutional reform in the developing countries. Rome: Food and Agriculture Organization of the United Nations, Extension, Education and Communication Service.

139. Robertson A (2012). Enabling agricultural extension for peach building, United States Institute of Peace, Washington.

140. Robinson RD (1988). The international transfer of technology: theory, issues, and practices. *Ballinger Publishing Company*.

141. Rochardson D (2003). Agricultural Extension Transforming ICT, Championing Universal Access. *Paper presented at the ICTs- Transforming Agricultural Extension?* CTA's observatory on ICTs, Wageningen.

142. Rodney Reynar, Fred Musser, Bruening T. The Potential For Linking Private And Public Extension Services In Bangladesh. *Journal of International Agricultural and Extension Education,* 1996;3:76.

143. Rogers EM (1962). *Diffusion of Innovations*. New York: Free Press.

144. Rogers EM (2003). *Diffusion of Innovations*, 5th Ed. New York: Free Press.

145. Roling N (1988). Extension Science: Information System in Agricultural Development. Cambridge: *Cambridge University Press*.

146. Roling N (1989). Misleading Metaphors. In D. Merrill-Sands, and D. Kaimowitz (Eds.). Technology Triangle: Linking Farmers, Technology Transfer Agents and Agricultural Researchers, *The Hague*: ISNAR.

147. Roling N. The research/extension interface: a knowledge system perspective. *ISNAR Staff Notes,* 1989; 48–89.

148. Rolling NG, Wagemakers MAE (Eds) (1998). Facilitating Sustainable Agriculture: Participatory Learning and Adaptive Management in Times of Environmental University, Cambridge: *Cambridge University Press*.

149. Rudra A (1982). Indian Agricultural Economics: Myths and Realities, New Delhi: *Allied Publishers Ltd.*

150. Rutton VW, Hayamid. Technology transfer and agricultural development. *Technology and Culture*, 1973;14(2):119–51.

152. Saha A (2011). Natural Resource Management – An Emerging Domain in Sustainable Development, In: D. Das Gupta (Ed.) (2011), Environmental Concerns, pp 127–32, ISBN No. (10): 81-7754-428-4. Agrobios, Jodhpur, India.

152. Samanta RK (1985). Technology Transfer and Appropriate Technology: *Paper presented in the summer institute on Agricultural Research management for Faculty of Agricultural University Management*. NAARM, Hyderabad.

153. Scoones L, Thompson J. (1994). Beyond Farmer First: Rural People 's Knowledge, Agricultural Research and Extension Practices. *London Intermediate Technology Publications*.

154. Shackleton S, Shackleton C, Cousins B (2000). Re-Valuing the Communal Lands of Southern Africa. New Understanding of Rural Livelihoods. Retrieved from ODI Natural Resource Perspective website: http:// www. odi.org.uk/nrp/62.html.

155. Sharma A (2012). Agriculture Knowledge Information System. Ph.D. Research Scholar Dept. of Agricultural Communication, GB Pant University of Ag.& Tech. Pantnagar, Uttarakhand.

156. Siegel DS, Waldman DA, Atwater LE, Link AN. Toward a model of the effective transfer of scientific knowledge from academicians to practitioners: qualitative evidence from the commercialization of university technologies. *Journal of Engineering and Technology Management*, 2004;21(1-2):115–42.

157. Singe H. (1978). Educational Conditions of Technology for Rural Development. In S Radhakrishna and MM Singh, *Technology for Rural Development (COSTED),* Bangalore, Indian Institute of Science.

158. Singh JP, Swanson BE, Singh KM. Developing a decentralized, market-driven extension system in India: The ATMA model. In AW van den Ban and RK Samanta, eds. changing roles of agricultural extension in Asian nations 2006;203–23.

159. Singh KM, Swanson BE. *Developing a market-driven extension system in India.* Annual Conference Proceedings of the Association for International Agricultural and Extension Education 2006;22:627–37.

160. Singh KM, Meena MS (2011). Agricultural Innovations in India-Experiences of ATMA Model. *Available at SSRN 1989823.*

161. Singh KM, Meena MS, Swanson BE, Reddy MN, Bahal R (2014). In-depth Study of the Pluralistic Agricultural Extension System in India. ICAR-RCER, Patna, UIUC, Illinois, MANAGE, Hyderabad, IARI, New Delhi.

162. Singh Kartar (2009). Rural Development: Principles, Policies and Management (3rd edition). *Sage Publication.*

163. Skolimowski H. The Structures of thinking in technology. *Technology and Culture.* 1966;7 (3):371–83.

164. Smith A (1759). The Theory of Moral Sentiments, Liberty Fund, Indiana Polis.

165. Smith A (1776). The Wealth of Nations, *The Modern Library*, New York.

166. Smith LD (1997). Decentralisation and Rural Development; Rome: FAO.

167. Sontakki BS, Samanta RK, Shenoy S, Reddy PV. Management Training Needs of Agricultural Scientists of Indian Council of Agricultural Research. Indian *Journal of Training and Development,* 2006;32(2):136–48.

168. Souder WE, Naskar AS, Padmanathan V. A guide to the best technology transfer practices. *Journal of Technology Transfer,* 1990;15(1–2):15–6.

169. Souder WE. (1987). Managing New Product innovation. Mass, New York: DC Health and Co.

170. Spivey WA, Munson JM, Nelson MA, Dietrich GB. Coordinating the technology transfer and transition of information technology: a phenomenological perspective. *IEEE Transactions on engineering management,* 1997;44(4):359–66.

171. Sulaiman VR, Hall A, Raina R (2006). "From Disseminating Technologies to Promoting Innovation: Implications for Agricultural Extension." Prepared for the SAIC Regional Workshop on Research–Extension Linkages for Effective Delivery of Agricultural Technologies in SAARC Countries, November 20–22. www.innovationstudies.org/docs/Rasheed-SAICNAARM-Paper.pdf, accessed December 2007.

172. Sulaiman RV (2012), "Agricultural Extension in India: Current Status and Way Forward," Background Paper prepared for the Round Table Consultation on Agricultural Extension, Beijing, Mar 15–17.

173. Sulaiman RV, Hall AJ. An innovation system perspective on the restructuring of agricultural extension: Evidence from India. *Outlook on Agriculture* 2002;30(4):235–43.

174. Sulaiman RV, Sadmate VV (2000). Privatizing agricultural extension in India. NCAP Policy Paper No. 10. New Delhi: National Center for Agricultural Economics and Policy Research.

175. Sulaiman V, Hall, A. (2004). Towards Extension-plus Opportunities and Challenges.

176. Suman RS. Attitude of farmers towards sustainability of vegetable cultivation. *Journal of AgriSearch* 2014;1(1):1–3.

177. Sung TK, Gibson DV. Knowledge and technology transfer grid: empirical assessment. *International Journal of Technology Management*, 2005;29(3-4):216–30.

178. Swanson BE. The changing role of agricultural extension in a global economy. *Journal of International Agricultural and Extension Education* 2006;13(3):5–17.

179. Swanson BE, Rajalahti R (2010). Strengthening agricultural extension and advisory systems.

180. Swanson BE, Bentz RP, Sofranko AJ (Eds.) (1997). Improving agricultural extension: A reference manual, Food and Agriculture Organization of the United Nations Rome.

181. Swanson B, Singh KM, Reddy MN (2008). A decentralized, participatory, market-driven extension system: The ATMA model in India. Participatory, Market-Driven Extension System: The ATMA Model in India (October 10, 2008).

182. Swanson BE. The changing role of agricultural extension in a global economy. *Journal of International Agricultural and Extension Education.* 2006;13(3):5–17.

183. Swanson BE (2008). Global Review of Good Agricultural Extension and Advisory Service Practices, Rome: United Nations Food and Agriculture Organization.

184. Swanson BE, Mathur PN (2003). Review of the Agricultural Extension System in India, the World Bank, July 2003.

185. Swanson BE, Samy MM (2004). Introduction to Decentralisation of Agricultural Extension Systems: Key elements for success. In W. Rivera, and Alex, G (Ed.). Decentralized Systems: Case Studies of International Initiatives. (pp. 1-5). Washington D.C.: The World Bank.

186. Swanson BE, Farner BJ, Bahal R (1990). The current status of agricultural extension worldwide. In BE Swanson (ed.) *Report of the Global Consultation on Agricultural Extension*, Rome, FAO.

187. Teffera Betru, Hamdar B. Strengthening the Linkages between Research And Extension In Agricultural Higher Education Institutions In Developing Countries. *International Educational Development*, 1997;17:303–11.

188. Tenkasi RV, Mohrman SA (1995). Technology transfer as collaborative learning. In TE Backer; SL David and G Soogy (Eds.). Reviewing the Behavioral Science Knowledge Base on Technology Transfer (pp. 147–67). Rockville, MD: U.S. Department of Health and Human Services, Public Health Service, National Institute of Health.

189. Thrupp, et al. (2000). Stakeholders and agents involved in the agricultural technology transfer model. In Anandajayasekeram P, Purkur R, Sindu W and Hockstra D (2008). *Concepts and practices in agricultural extension in developing countries: A source book.* IFPRI, Washington DC, USA and ILRI, Nairobi, Kenya. 275 pp.

190. Toness A. The potential of Participatory Rural Appraisal (PRA) approaches and methods for agricultural extension and development in 21st century. *Journal of International Agriculture and Extension Education,* 2001;8:25–37.

191. United Nations Economic and Social Council. (2003). *Promoting an Integrated Approach to Rural Development in Developing Countries for Poverty Eradication and Sustainable Development.* Report of the Secretary General. New York: Economic and Social Council. (www.un.org/esa/coordination/Alliance/documents/website/ E-2003-51(eng).pdf).

192. United Nations Environment Programme (2007a). Global Environmental Outlook: Environment for Development. Nairobi: United Nations Environment Programme. (www.unep.org/geo/geo4/media).

193. Van Crowde L (1996). Agricultural Extension for Sustainable Development. SD Dimensions, Rome: FAO. Retrieved from FAO website http:// www/ fao.org/sd EXdirect/ EXan004.htm.

194. Van den Ban AW, Sulaiman Rasheed V. (2000). Agricultural Extension in India- The Next Step. *Policy Brief* (9), NCAP, ICAR.

195. Van Den Ban A. Different Ways of Financing Agricultural Extension. *AgREN Network Paper,* 2000;106:8–19.

196. Van Den Ban A, Wageningen N. Funding and Delivering Agricultural Extension in India. Rasheed Sulaiman V. *Journal of International Agricultural and Extension Education,* 2003;10:21.

197. Van den Ban AW, Hawkins BS (1998). Agricultural Extension. *CBS Publishers and Distributors,* New Delhi.

198. Van den Ban AW, Hawkins HS (1988). Agricultural Extension. *Burnt Mill,* Harlow.

199. Van den Berg H. (2004). *IPM Farmer Field Schools: A Synthesis of 25 Impact Evaluations.* Prepared for the Global IPM Facility of Wageningen University, Netherlands. (www.fao.org/docrep/006/ad487e/ad487e00.htm).

200. Venkatasubramanian V, Sajeev MV, Singha AK (2009). Concept, Approaches and Methodologists for Technology Application and Transfer-A Resource Book for KVKs. Meghalaya, India: Zonal Project Directorate Zone III, ICAR.

201. VISION 2020—Indian Council of Agricultural Research, India.

202. WAGE (Working Group on Agricultural Extension) (2007). Recommendations of Working Group on Agricultural Extension for Formulation of 11th Five Year Plan (2007-12). New Delhi Planning Commission, Government of India.

203. Waghmare SK, Pandit VK. Constraints in Adoption of wheat technology by the tribal farmers. Indian Journal of Extension Education. 1982;XVIII (1 &2):95–8.

204. World Bank (2012). Agricultural Innovations System—An Investment Source Book. Washington DC: The World Bank.

205. World Bank, (2000a). Decentralising Agricultural Extension: Lesson and Good Practices. Retrieved from website: http:// siteresources. World bank.org/INTARD/ 8258261110 63678817/20431788/ Decentralisation. Pdf.

206. World Bank (1994). Agricultural Extension in Africa: African Technical Development Series. Washington DC, USA: The World Bank.

207. World Bank (1996a). Participatory rural appraisal (appendix I). In *The World Bank Participation Sourcebook.* Washington, DC: World Bank. (www.worldbank.org/wbi/sourcebook/sba104.htm).

208. World Bank (2006). World development indicators. Washington DC.

209. World Bank. (2006b). Enhancing Agricultural Innovation: How to Go Beyond the Strengthening of Research Systems. Washington, DC: Agricultural and Rural Development, World Bank.

210. World Bank (2007). *Agriculture for development. World development report 2008.* Washington DC: The International Bank for Reconstruction and Development/World Bank.

211. World Bank. (2004b). *Making services work for poor people. World development report 2004.* Washington DC: World Bank and Oxford University Press.

212. Yotopoulos PA, Nugent JB (1976). Economics and Development: Empirical Investigations. New York: *Harper and Row Publication.*

213. Yudelman M. Agriculture in integrated rural development: The experience of the World Bank. *Food Policy* 1976;1(5):367–81. (http://dx.doi.org/10.1016/0306- 9192(76)90072-5).

Index

Reader's Notes

Reader's Notes

Reader's Notes

Reader's Notes

Reader's Notes

Reader's Notes